Advances in Industrial Control

Other titles published in this series:

Hassan Noura • Didier Theilliol
Jean-Christophe Ponsart • Abbas Chamseddine

Fault-tolerant Control Systems

Design and Practical Applications

 Springer

Hassan Noura, PhD
United Arab Emirates University
Department of Electrical Engineering
P.O. Box 17555
Al-Ain
UAE
hnoura@uaeu.ac.ae

Abbas Chamseddine, PhD
Université Paul Cézanne, Aix Marseille III
Domaine Universitaire de Saint-Jérôme
Laboratoire des Sciences de l'Information
et des Systèmes
Avenue Escadrille Normandie-Niemen
13397 Marseille cedex 20
France
abbas.chamseddine@lsis.org

Jean-Christophe Ponsart, PhD
Nancy-Université
Faculté des Sciences et Techniques
Centre de Recherche en Automatique
de Nancy CNRS UMR 7039
BP 70239
54506 Vandoeuvre cedex
France
jean-christophe.ponsart@cran.uhp-nancy.fr

Didier Theilliol, PhD
Nancy-Université
Faculté des Sciences et Techniques
Centre de Recherche en Automatique
de Nancy CNRS UMR 7039
BP 70239
54506 Vandoeuvre cedex
France
didier.theilliol@cran.uhp-nancy.fr

ISSN 1430-9491
ISBN 978-1-4471-2671-3   e-ISBN 978-1-84882-653-3
DOI 10.1007/978-1-84882-653-3
Springer Dordrecht Heidelberg London New York

British Library Cataloguing in Publication Data
A catalogue record for this book is available from the British Library

Cover design: eStudioCalamar, Figueres/Berlin

Printed on acid-free paper

Springer is part of Springer Science+Business Media (www.springer.com)

Advances in Industrial Control

Professor (Emeritus) O.P. Malik
Department of Electrical and Computer Engineering
University of Calgary
2500, University Drive, NW
Calgary, Alberta
T2N 1N4
Canada

Professor K.-F. Man
Electronic Engineering Department
City University of Hong Kong
Tat Chee Avenue
Kowloon
Hong Kong

Professor G. Olsson
Department of Industrial Electrical Engineering and Automation
Lund Institute of Technology
Box 118
S-221 00 Lund
Sweden

Professor A. Ray
Department of Mechanical Engineering
Pennsylvania State University
0329 Reber Building
University Park
PA 16802
USA

Professor D.E. Seborg
Chemical Engineering
3335 Engineering II
University of California Santa Barbara
Santa Barbara
CA 93106
USA

Doctor K.K. Tan
Department of Electrical and Computer Engineering
National University of Singapore
4 Engineering Drive 3
Singapore 117576

Professor I. Yamamoto
Department of Mechanical Systems and Environmental Engineering
The University of Kitakyushu
Faculty of Environmental Engineering
1-1, Hibikino,Wakamatsu-ku, Kitakyushu, Fukuoka, 808-0135
Japan

To our families and parents

Series Editors' Foreword

The series *Advances in Industrial Control* aims to report and encourage technology transfer in control engineering. The rapid development of control technology has an impact on all areas of the control discipline. New theory, new controllers, actuators, sensors, new industrial processes, computer methods, new applications, new philosophies..., new challenges. Much of this development work resides in industrial reports, feasibility study papers, and the reports of advanced collaborative projects. The series offers an opportunity for researchers to present an extended exposition of such new work in all aspects of industrial control for wider and rapid dissemination.

Control system design and technology continues to develop in many different directions. One theme that the *Advances in Industrial Control* series is following is the application of nonlinear control design methods, and the series has some interesting new commissions in progress. However, another theme of interest is how to endow the industrial controller with the ability to overcome faults and process degradation. Fault detection and isolation is a broad field with a research literature spanning several decades. This topic deals with three questions:

- How is the presence of a fault detected?
- What is the cause of the fault?
- Where is it located?

However, there has been less focus on the question of how to use the control system to accommodate and overcome the performance deterioration caused by the identified sensor or actuator fault.

One approach to all these issues is to institute a rigorous programme of process and controller monitoring to minimize the actual occurrence of fault situations altogether, so-called preventative maintenance. The response of the control community to these questions has been a little diffuse with no method promoted as a clear winner; for example, for a period the method of "reliable control" featured in the literature and at control conferences. This approach relaxes the control to accommodate fault failures but the cost is degraded performance. Another approach was that of "reconfigurable control" in which the controller is restructured to use the remaining available sensors and actuators and to use any available knowledge of the changes that have occurred to the process or system.

However, from this diversity of approaches, the industrial control engineer or the process engineer is quite likely to ask "But do any of these methods work and what is the online industrial overhead for the application of these methods?" One way to find out is to study this new volume *Fault-tolerant Control Systems: Design and Practical Applications* by Hassan Noura, Didier Theilliol, Jean-Christophe Ponsart, and Abbas Chamseddine. This monograph is a little different from some in the *Advances in Industrial Control* series, for it is a very focused study of the application of fault-tolerant control (FTC) to three increasingly complex processes. It has the objective of showing the reader all the steps in design, implementation, and assessment for some laboratory-scale and industrial applications of FTC.

The monograph opens with a brief introductory chapter, but it is in Chap. 2 where all the components of FTC design are presented. Six sections of the chapter build up the design systematically by introducing process models (linear and nonlinear), fault descriptions (both actuators and sensors), the nominal tracking control design, fault diagnosis methods (using residual generation), fault effect and size estimation and controller compensation mechanisms to accommodate the fault. All this is brought together and presented in the final section as a generic FTC system architecture. Three extended case-study chapters, which give this monograph its distinctive character, follow this focussed presentation of the methods to be used. The three process applications are a laboratory-scale winding machine that characterises "simple" complexity, a laboratory-scale three-tank process that represents "middle-level" complexity, and finally "complex" complexity is represented by an industry-grade automotive active suspension system. The value of these case studies lies in their fully documented thoroughness that gives the reader a good practical insight into how the methods work and allows the possibility of replicating the three studies themselves.

The practical emphasis of the monograph and its case studies will appeal to a wide range of academic researchers and industrial control and process engineers. Academics and students will be able to repeat the case studies on in-house laboratory equipment, whilst the industrial engineer should obtain a better insight as to how FTC can be implemented on industrial processes.

This is a growing field and the *Advances in Industrial Control* series has a number of volumes available on FTC and related subjects. It is perhaps worth mentioning *Soft Sensors for Monitoring and Control of Industrial Processes* by Luigi Fortuna and colleagues (ISBN 978-1-84628-479-3, 2007), *Process Control Performance Assessment* edited by Andrzej W. Ordys and colleagues (ISBN 978-1-84628-623-0, 2007), *Diagnosis of Process Nonlinearities and Valve Stiction* by M.A.A. Shoukat Choudhury and colleagues

(ISBN 978-3-540-79223-9, 2008), and finally Guillaume Ducard's monograph *Fault-tolerant Flight Control and Guidance Systems* (ISBN 978-1-84882-560-4, 2009). Clearly, the *Advances in Industrial Control* series has some very strong entries to this growing technical field and the editors are pleased to see the monograph *Fault-tolerant Control Systems* by Hassan Noura and colleagues join the collection.

Industrial Control Centre   *M.J. Grimble*
Glasgow, Scotland, UK   *M.A. Johnson*
2009

Foreword

As technological systems become more and more complex, the dependence on their control systems has also increased significantly. This is particularly true in safety-critical applications where either the success of a mission or ultimate protection of human lives, property, and environment becomes a paramount goal. For any practical control systems, no matter how ingenious the design is, and how immaculate the manufacture process is carried out, things will break. It is a matter of time, sooner or later. One way to ensure reliable operation of a system for intended purposes, despite those undesirable circumstances, such as failures, is to rely on fault-tolerant control strategies.

A fault-tolerant control system is a control system specifically designed with potential system component failures in mind. Clearly, a fault-tolerant control may not offer optimal performance in a strict sense for normal system operation, but generally it can mitigate effects of system component failures without completely jeopardizing the mission or putting the users/public at risk. Clearly, the philosophy of fault-tolerant control systems design is different from other design methodologies. Consequently, their behavior under system component failures will also be different.

Design of control systems to achieve fault-tolerance for closed-loop control of safety-critical systems has been an active area of investigation for many years. It becomes more and more clear that there are certain trade-offs between achievable normal performance and fault-tolerance capability. A fault-tolerant control system design has essentially become a decision on manipulation of such trade-offs.

Despite the efforts in control system community, the field of fault-tolerant control systems is still wide open. Most of the contributions so far are theoretical in nature. It is important to emphasize that when a failure occurs in a system, either in sensors or actuators, the characteristics of the entire system can undergo significant change, i.e., degradation. The actuators may not be able to provide the same level of driving power, while the sensors may not supply dependable measurements. Without full understanding of those practical constraints and respecting the failure induced limitations, fault-tolerant control system design based purely on theory will be bound to fail in practice.

In the past decade, most of the work in this area has been theoretical in nature. It is refreshing to read this book which has put the emphasis on the practical applications. This book will certainly be an important addition to the library on fault-tolerant control systems. It is one of very few books in this area that considers practical aspects of fault-tolerant control. It is certainly a welcome addition and valuable reference for anyone working in this area.

The basic concepts of fault-tolerant control systems are introduced in Chap. 1. Classification of fault-tolerant control strategies is presented in terms of fault severity levels. Fault-tolerant control system design and analysis against actuator and sensor failures have been treated in detail in Chap. 2 for linear and nonlinear systems. All the important concepts have been presented using physical system examples by comparing normal system performance against those under component failures. Both partial and complete failures of sensors and actuators have been considered. In this chapter, techniques for fault diagnosis and fault estimations have also been presented. Finally, a general architecture of a fault-tolerant control system is developed.

Chapter 3 is devoted to the application of fault-tolerant control strategies on a physical lab-scale winding machine. The authors have provided basic control objectives for this system with sufficient detail. The performance of the system under normal conditions is analyzed first to provide a baseline benchmark for fault-tolerant control system design and analysis. Subsequently, various actuator and sensor failure scenarios have been dealt with. In this chapter, both linearized and nonlinear system models are considered. The book has clearly shown that the effects of faults can be compensated with properly designed fault-tolerant control systems.

A well known three-tank system is used in Chap. 4 to illustrate design and analysis techniques for fault-tolerant control systems. An advantage of choosing this system is that the physical relationships among key system variables can be easily obtained. Using the dynamic models obtained, the detailed procedure for fault-tolerant control system design and analysis can be clearly demonstrated. The authors have also included some MATLAB® scripts to guide the readers through the process and to encourage readers to try by themselves. Both linear and nonlinear system models have been utilized in the design and analysis process.

Finally, fault-tolerant control system design and analysis against sensor failures in an active suspension of a full vehicle system have been considered. Detailed mathematical description of the suspension is derived first. Based on this model, fault-tolerant control systems performance against several commonly encountered sensor failures have been investigated. The originality of the work in this chapter is the breakdown of the entire suspension system into several interconnected subsystems. Each subsystem has its own local controller and its own fault diagnostic module. A higher level control system coordinates the information issued from these local modules.

In summary, the authors have successfully presented some most important concepts and procedures in fault-tolerant control system design and analysis. The authors have done this with elegance of mathematics, as well as in-depth physical understanding of the limitations of handicapped actuators and sensors. This is a must read book on the subject of fault-tolerant control systems.

The logical introduction and the easy to understand styles of presentation have made this book particularly suitable for graduate students and practising engineers who are looking for some guidance in applying active fault-tolerant control methods in their own fields of interests.

London, Ontario, Canada,   *Professor Jin Jiang*
March 2009

Acknowledgements

We want to thank our colleagues who have been working with us for many years of research in FDI and FTC topics. Our special thanks go to Professor D. Sauter for purchasing the winding machine and the three-tank system to be used in teaching and research.

We would also like to thank Master and PhD students who contributed to this book through their work in different ways: M. Adam-Medina, T. Bastogne, V. Begotto, V. Dardinier, C. Gaubert, J-M. Hitinger, C. Join and M. Rodrigues. Our thanks also go to Dr. H. Hejase and R. Hejase for their useful comments and corrections.

Contents

1

Introduction to Model-based Fault Diagnosis and Fault-tolerant Control

The automation of a process consists in providing a quasi-optimal solution to obtain the best possible quality of the final product and consequently an increase in profits. Automated system control theory has been widely developed and applied to industrial processes. These techniques ensure the stability of the closed-loop system and yield a pre-defined performance in the case where all system components operate safely. However, the more the process is automated, the more it is subject to the occurrence of faults. Consequently, a conventional feedback control design may result in an unsatisfactory performance in the event of malfunctions in the actuators, sensors, or other components of the system. This may even lead the system to instability. In highly automated industrial systems where maintenance or repair cannot always be achieved immediately, it is convenient to design control methods capable of ensuring nominal performance when taking into account the occurrence of faults. This control is referred to as fault-tolerant control (FTC) which has become of paramount importance in the last few decades. The design of an FTC system requires obviously quick fault detection and isolation (FDI) for adequate decision making. Hence, to preserve the safety of operators and the reliability of processes, the presence of faults must be taken into account during the system control design.

1.1 Fault Diagnosis

Process monitoring is necessary to ensure the effectiveness of process control and consequently a safe and a profitable plant operation. Sensor or actuator failure, equipment fouling, feedstock variations, product changes, and seasonal influences may affect the controller performance. Such issues apparently make up to 60% of industrial controllers problems [61]. FDI refers to the task of inferring the occurrence of faults in a process and finding their root causes using the following various knowledge-based system strategies: quantitative models [128], qualitative models [126], and historical data [127].

The diagnosis of such problems from raw data trends is often difficult. However, quantitative or qualitative model-based FDI techniques are considered and combined to supervise the process and to ensure appropriate reliability and safety in industry. Short historical review of model-based FDI can be found in [43, 70, 86] and current developments are reviewed in [44]. Among quantitative models, FDI for linear/nonlinear systems remains a challenge due to the problem of discriminating between disturbances and faults through a wide range of operating conditions. Model-based FDI methods have been developed for exact and uncertain linear/nonlinear mathematical description of systems based on observer schemes, parameter estimation algorithms, or parity space techniques. Several books are dedicated to these topics such as [22,24,53] or more recently [31]. For fault isolation, various techniques based on an exact knowledge of the nonlinear model allow us to generate residuals sensitive to specific faults and insensitive to others using decoupling methods [49,80] or geometric approaches [60,101].

Based on the large diversity of advanced model-based methods for automated FDI, the problem of actuator or/and sensor fault detection (which is one of the main targets of this book) is of basic importance. Nevertheless, due to difficulties inherent in the on-line identification of closed-loop systems, parameter estimation techniques are not always suitable. The parity space technique is suitable to distinguish between different faults in the presence of uncertain parameters, but is not useful for fault magnitude estimation. However, the observer-based method is more appropriate to achieve this objective. Classical decoupled techniques, such as unknown input observer or dedicated filter devoted to detect and estimate faults (considered as unknown inputs), can be synthesized for solving the FDI problem in certain cases. It should be highlighted that the fault magnitude estimation is necessary to ensure an accurate fault monitoring in order to provide an efficient maintenance operation and to ensure the safety of the environment.

Moreover, many accidents with airplanes or in nuclear power plants have dramatically illustrated the very important step of FDI to inform the operators about the system's status. However, it appears clear that detecting and isolating a fault is not sufficient if no subsequent action occurs once a fault has been identified.

As recently proposed by [14,59], an FTC system, based on fault isolation and magnitude estimation, can be envisioned to maintain control objectives despite the occurrence of a fault.

1.2 Fault-tolerant Control

Much effort has been made in the field of FTC in the presence of faults in the functioning of the nuclear and avionics industries, chemical or petrochemical plants, etc. The various studies dealing with this problem are based on hardware or analytical redundancy. Hardware redundancy is necessary in systems where the safety of people could be affected (airplanes, nuclear plants).

In other industrial processes, hardware redundancy is rare or non-existant, because of its expensive financial cost. Redundant sensors, usually much easier and less expensive than actuators, are generally installed. Thus, in the presence of a major actuator failure, it is impossible to maintain the damaged system at some acceptable level of performance. It becomes of prime importance to lead it to its optimal operating order, with respect to desirable performances and degree of priority. Therefore, the main feature is to minimize the loss in productivity (lower quality production) or/and to operate safely without danger to human operators or equipment. The system can continue its operation with decreased performance as long as it remains within acceptable limits. The use of analytical redundancy makes possible the reduction in instruments cost and maintenance.

The topic of fault-tolerance has attracted the interest of many researchers worldwide. Recently, a very interesting bibliographical review of fault-tolerance was performed by Zhang and Jiang [140]. Various books on FTC have also been published recently [13, 14, 17, 59, 69, 92].

The analytical fault-tolerant operation can be achieved either passively by the use of a control law designed to be insensitive to some known faults, or actively by an FDI mechanism, and the redesign of a new control law. The active methods are more realistic because all the faults that may affect the system cannot be known *a priori*.

Unlike the fault diagnostic field, where definitions and classification of methods have been clearly given in the literature, FTC is still missing standard definitions and classifications. In this book we propose a classification of FTC techniques. This classification is illustrated by the control system performance vs the severity of the failure (Fig. 1.1). As previously stated, for *a priori* known faults, a controller with fixed parameters could be set up with the objective of controlling the nominal system as well as the system affected by these known faults (passive methods). This strategy can be achieved using the simultaneous stabilization methods or methods based on robust control (H_∞, disturbance rejection, *etc.*). These techniques are also known as *reliable control* techniques where the controller must be insensitive to the occurrence of specific faults.

However, it is obvious that passive methods are very restrictive because all the expected faults and their effects on the plant cannot be known *a priori*. Active approaches are preferable to deal with an increasing number of faults. These methods consist of adjusting the controllers on-line according to the fault magnitude and type, in order to maintain the closed-loop performance of the system. If it remains possible to preserve the faulty system performance close to the nominal one, active methods correspond to *reconfiguration*. For more critical failures (a complete loss of an actuator), the nominal performance cannot be maintained anymore, the current performance are reduced as shown in the shaded area of Fig. 1.1. In such cases, a *restructuring* strategy consisting of modifying the system structure or the control objectives is

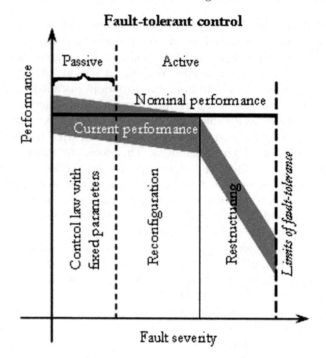

Fig. 1.1. FTC strategies

used. For instance, the number of controlled outputs has to be reduced, or the nominal reference inputs cannot be reached anymore, and other references have to be redefined. The objective is to lead the system into a degraded operating mode. Moreover, for a certain type of failure, it is impossible to keep the system operating even in degraded mode. In this case, the aim is to shut down the system safely.

In general, an FTC strategy must include an FDI module and an upper level of supervision module aiming at:

- Dealing with various kinds of faults affecting the plant
- Providing information about the system behavior, the degradations produced by the fault and the new performance reached by the degraded system
- Deciding if an FTC method has to be switched on, or if the system must shut down

In the literature, FTC methods are developed considering that the FDI is already achieved. FDI techniques are rarely integrated into FTC systems. It is very often assumed that the model of the faulty system is known, which is not always realistic for unexpected faults. In order to eliminate the actuator fault effect which occurs on the system, various methods have been proposed to recover, as close as possible, the performance of the pre-fault system according to the considered fault representation. Two main approaches have been developed. One is based on a model matching principle where the control gain is completely re-computed on-line, and the other method is based on adding fault compensation to the nominal control law. In the first approach, the control system can be designed so that the faulty system performance is recovered and the new system behaves as originally specified. Gao and Antsaklis [47] suggest a basic approach based on what they called the pseudo-inverse method (PIM). Rather than the *exact model matching* method proposed by [48, 117] has recently proposed an extended PIM method to develop an *admissible model matching* approach. These gain redesign methods are not considered in this book which is completely dedicated to the method based on the additive control law. From a general point of view, one objective of this book is to show the development of complete FTC methods where the control law is modified once a fault has been detected, isolated, and estimated. Another objective is to show the application of these methods to real laboratory-scale systems taking into account the constraints of the real applications.

In Chap. 2, the study is developed showing the various steps of active methods according to the severity of the faults and the ability of the system to tolerate them. First, actuators and sensors are considered to operate badly but they are still able to achieve part of the original system performance. Then a complete loss of a sensor and an actuator are analyzed. The complete loss of an actuator is a major failure which leads to a large decrease in system performance. Thus, it is necessary to restructure the control objectives with degraded performance. It is shown that, according to the classification described previously, the strategy to adopt and the level of performance to recover depend on:

- The process itself
- The degree of the available (hardware and/or analytical) redundancy in the system
- The severity of the fault or the failure
- The level of desired performance

In this chapter, linear and nonlinear systems are considered. A nominal control law is described for both cases. For the case of linear or linearized systems, the notion of an operating point is detailed. Actually, this notion is not always very well understood and needs to be highlighted. Nominal tracking control and model-based fault diagnostic techniques are described in detail and used in the design of the FTC. These methods will be illustrated by applying them to different systems as will be detailed in the following chapters.

Chapter 3 is dedicated to the application of the FTC methods described in Chap. 2 to a laboratory-scale system representing a winding machine. This electro-mechanical system is a nonlinear system and its model and parameters are not easy to obtain. Therefore, a model linearized around an operating point is obtained experimentally as a black box model. Later, a multiple model technique is described to allow the system to range over the whole operating region.

In Chap. 4, the well known three-tank system is used to illustrate FTC methods and results. For this system, the physical equations are easily obtained allowing us to write a nonlinear model of the system. This model is first linearized around an operating point and used in the design and the application of FDI and FTC methods. Later in this chapter, the nonlinear model is used to deal with the case of major actuator faults such as actuator loss. In this case, the system is driven outside the linearized zone and the linear model is no longer valid, yielding a bad performance of the control system.

Chapter 5 considers a nonlinear complex system represented by a car active suspension. Unfortunately, a real active suspension system is not available, but the physical model is described according to papers from the literature and validated on real systems. This system is considered to be complex due to its large number of variables. The originality of the work presented in this chapter is in the breakdown of this system to interconnected subsystems. Each subsystem has its own local controller and its own fault diagnostic module. A higher level module coordinates the information issued from these local modules. Simulation results are performed to illustrate the performance of an FTC method in the presence of sensor faults for complex systems. Moreover, as the number of state variables is large, the number of sensors to use for such systems is analyzed and existing sensors in industry are discussed.

Actuator and Sensor Fault-tolerant Control Design

2.1 Introduction

Many industrial systems are complex and nonlinear. When it is not easy to deal with the nonlinear models, systems are usually described by linear or linearized models around operating points. This notion of operating point is very important when a linearized model is considered, but it is not always easily understood. The objective in this chapter is to highlight the way to seek an operating point and to show a complete procedure which includes the identification step, the design of the control law, the FDI, and the FTC. In addition to the detailed approach dealing with linearized systems around an operation point, a nonlinear approach will be presented.

2.2 Plant Models

2.2.1 Nonlinear Model

Many dynamical system can be described either in continuous-time domain by differential equations:

$$\begin{cases} \dot{x}(t) = f(x(t), u(t)) \\ y(t) = h(x(t), u(t)) \end{cases},\qquad(2.1)$$

or in discrete-time domain by recursive equations:

$$\begin{cases} x(k+1) = f(x(k), u(k)) \\ y(k) = h(x(k), u(k)) \end{cases},\qquad(2.2)$$

where $x \in \Re^n$ is the state vector, $u \in \Re^m$ is the control input vector, and $y \in \Re^q$ is the system output vector. f and h are nonlinear functions.

These forms are much more general than their standard linear counterparts which are described in the next section.

There is a particular class of nonlinear systems – named input-linear or affine systems – which is often considered, as many real systems can be described by these equations:

$$\begin{cases} \dot{x}(t) = f(x(t)) + \sum_{j=1}^{m} (g_j(x(t))u_j(t)) \\ \qquad = f(x(t)) + G(x(t))u(t) \\ y_i(t) = h_i(x(t)), \qquad\qquad 1 \le i \le q \end{cases} \qquad (2.3)$$

$f(x)$ and $g_j(x)$ can be represented in the form of n-dimensional vector of real-valued functions of the real variables $x_1, \ldots, x_n$, namely

$$f(x) = \begin{bmatrix} f_1(x_1, \ldots, x_n) \\ f_2(x_1, \ldots, x_n) \\ \vdots \\ f_n(x_1, \ldots, x_n) \end{bmatrix} ; \qquad g_j(x) = \begin{bmatrix} g_{1j}(x_1, \ldots, x_n) \\ g_{2j}(x_1, \ldots, x_n) \\ \vdots \\ g_{nj}(x_1, \ldots, x_n) \end{bmatrix}. \qquad (2.4)$$

Functions $h_1, \ldots, h_q$ which characterize the output equation of system (2.3) may be represented in the form

$$h_i(x) = h_i(x_1, \ldots, x_n). \qquad (2.5)$$

The corresponding discrete-time representation is

$$\begin{cases} x(k+1) = f_d(x(k)) + \sum_{i=j}^{m} (g_{d_j}(x(k))u_j(k)) \\ \qquad = f_d(x(k)) + G_d(x(k))u(k) \\ y(k) = h_d(x(k)) \end{cases} \qquad (2.6)$$

2.2.2 Linear Model: Operating Point

An operating point is usually defined as an equilibrium point. It has to be chosen first when one has to linearize a system. The obtained linearized model corresponds to the relationship between the *variation of the system output* and the *variation of the system input* around this operating point. Let us consider a system associated with its actuators and sensors, with the whole range of the operating zone of its inputs U and measurements Y (Fig. 2.1).

If the system is linearized around an operating point (U_0, Y_0), the linearized model corresponds to the relationship between the variations of the system inputs u and outputs y (Fig. 2.2) such that

$$u = U - U_0 \qquad \text{and} \qquad y = Y - Y_0. \qquad (2.7)$$

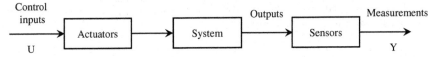

Fig. 2.1. System representation considering the whole operating zone

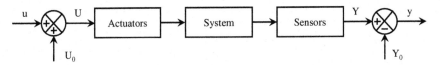

Fig. 2.2. System representation taking into account the operating point

Then the model describing the relationship between the input u and the output y can be given by a Laplace transfer function for single-input single-output (SISO) systems:

$$\Theta(s) = \frac{y(s)}{u(s)}, \qquad (2.8)$$

or by a state-space representation given in continuous-time for SISO or multiple-input multiple-output (MIMO) systems:

$$\begin{cases} \dot{x}(t) = Ax(t) + Bu(t) \\ y(t) = Cx(t) + Du(t) \end{cases}, \qquad (2.9)$$

where $x \in \Re^n$ is the state vector, $u \in \Re^m$ is the control input vector, and $y \in \Re^q$ is the output vector. A, B, C, and D are matrices of appropriate dimensions.

Very often, in real applications where a digital processor is used (microcontroller, programmable logic controller, computer, and data acquisition board, *etc.*), it may be more convenient to consider a discrete-time representation:

$$\begin{cases} x(k+1) = A_d x(k) + B_d u(k) \\ y(k) = C_d x(k) + D_d u(k) \end{cases}. \qquad (2.10)$$

A_d, B_d, C_d, and D_d are the matrices of the discrete-time system of appropriate dimensions.

In the sequel, linear systems will be described in discrete-time, whereas nonlinear systems will be considered in continuous-time. For the simplicity of notation and without loss of generality, matrix D_d is taken as a zero matrix, and the subscript d is removed.

2.2.3 Example: Linearization Around an Operating Point

To illustrate the notion of the operating point, let us consider the following example. In the tank presented in Fig. 2.3, the objective is to study the behavior of the water level L and the outlet water temperature T_o. An inlet flow rate Q_i is feeding the tank. An electrical power P_u is applied to an electrical resistor to heat the water in the tank.

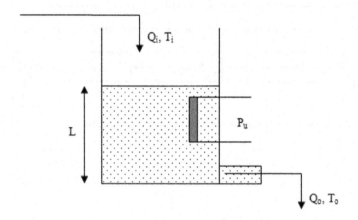

Fig. 2.3. Tank with heater

- Q_i is the inlet water flow rate
- T_i is the inlet water temperature considered as constant
- Q_o is the outlet water flow rate
- T_o is the outlet water temperature
- L is the water level in the tank
- P_u is the power applied to the electrical resistor
- S is the cross section of the tank

The outputs of this MIMO system are L and T_o. The control inputs are Q_i and P_u. The block diagram of this system is given in Fig. 2.4.

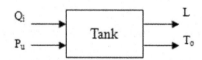

Fig. 2.4. The input/output block diagram

Assuming that the outlet flow rate Q_o is proportional to the square root of the water level in the tank ($Q_o = \alpha\sqrt{L}$), the water level L will be given by the following nonlinear differential equation:

$$\frac{dL(t)}{dt} = \frac{1}{S}(Q_i(t) - Q_o(t)) = \frac{1}{S}(Q_i(t) - \alpha\sqrt{L(t)}). \qquad (2.11)$$

Based on the thermodynamics equations, the outlet water temperature T_o is described by the following nonlinear differential equation:

$$\frac{dT_o(t)}{dt} = \frac{P_u(t)}{SL(t)\mu c} - \frac{T_o(t) - T_i(t)}{SL(t)} Q_i(t), \tag{2.12}$$

where c is the specific heat capacity, and μ is the density of the water.

The objective now is to linearize these equations around a given operating point: $OP = (Q_{i0}, P_{u0}, Q_{o0}, T_{o0}, L_0)$. Around this operating point, the system variables can be considered as

$$\begin{aligned} Q_i(t) &= Q_{i0} + q_i(t); \quad P_u(t) = P_{u0} + p_u(t); \quad Q_o(t) = Q_{o0} + q_o(t); \\ T_o(t) &= T_{o0} + t_o(t); \quad L(t) = L_0 + l_o(t). \end{aligned} \tag{2.13}$$

The linearization of (2.11) around the operating point OP is given by

$$\frac{dl(t)}{dt} = \frac{1}{S} q_i(t) - \frac{\alpha}{2S\sqrt{L_0}} l(t). \tag{2.14}$$

Similarly, the linearization of (2.12) around the operating point OP is given by

$$\begin{aligned} \frac{dt_o(t)}{dt} = &-\frac{T_{o0} - T_i}{SL_0} q_i(t) + \frac{1}{SL_0\mu c} p_u(t) \\ &- \frac{1}{L_0^2} \left(\frac{P_{u0}}{S\mu c} - \frac{T_{o0} - T_i}{S} Q_{i0} \right) l(t) \\ &- \frac{Q_{i0}}{SL_0} t_o(t). \end{aligned} \tag{2.15}$$

Considering the following state vector $x = \begin{bmatrix} l & t_o \end{bmatrix}^T$, the linearized state-space representation of this system around the operating point is then given by

$$\begin{cases} \dot{x}(t) = \begin{bmatrix} \dot{l}(t) \\ \dot{t}_o(t) \end{bmatrix} = \begin{bmatrix} -\dfrac{\alpha}{2S\sqrt{L_0}} & 0 \\ a & b \end{bmatrix} \begin{bmatrix} l(t) \\ t_o(t) \end{bmatrix} + \begin{bmatrix} \dfrac{1}{S} & 0 \\ c & d \end{bmatrix} \begin{bmatrix} q_i(t) \\ p_u(t) \end{bmatrix}, \\ y(t) = \begin{bmatrix} l(t) \\ t_o(t) \end{bmatrix} = \begin{bmatrix} 1 & 0 \\ 0 & 1 \end{bmatrix} \begin{bmatrix} l(t) \\ t_o(t) \end{bmatrix} \end{cases} \tag{2.16}$$

where

$$a = -\frac{1}{L_0^2} \left(\frac{P_{u0}}{S\mu c} - \frac{T_{o0} - T_i}{S} Q_{i0} \right); \qquad b = -\frac{Q_{i0}}{SL_0};$$

$$c = -\frac{T_{o0} - T_i}{SL_0}; \qquad d = \frac{1}{SL_0\mu c}.$$

Numerical Application

Study the response of the system to the *variation of the input variables* as follows: $q_i = 10 \; l/h$ and $p_u = 2 \; kW$. The numerical values of the system lead to the following state-space representation:

$$\dot{x}(t) = \begin{bmatrix} \dot{l}(t) \\ \dot{t}_o(t) \end{bmatrix} = \begin{bmatrix} -0.01 & 0 \\ 0 & -0.02 \end{bmatrix} \begin{bmatrix} l(t) \\ t_o(t) \end{bmatrix} + \begin{bmatrix} 0.01 & 0 \\ -0.03 & 0.04 \end{bmatrix} \begin{bmatrix} q_i(t) \\ p_u(t) \end{bmatrix}. \quad (2.17)$$

The simulation results of the linearized system in response to the variation of the system inputs in open-loop around the operating point are shown in Fig. 2.5.

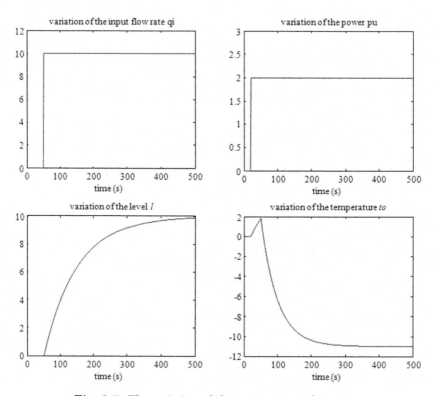

Fig. 2.5. The variation of the system inputs/outputs

It can be seen that initial values of these variables are zero. The zero here corresponds to the value of the operating point. However, the real variables Q_i, P_u, L, and T_o are shown in Fig. 2.6.

Later on, if a state-feedback control has to be designed, it should be based on the linearized equations given by (2.16).

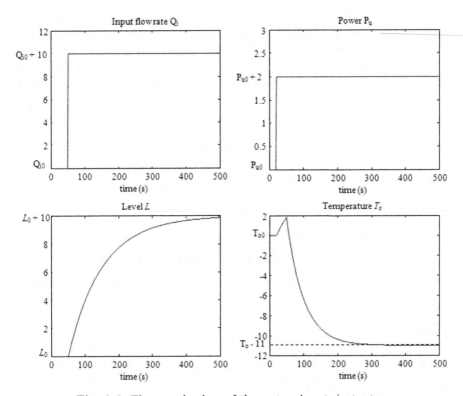

Fig. 2.6. The actual values of the system inputs/outputs

2.3 Fault Description

During the system operation, faults or failures may affect the sensors, the actuators, or the system components. These faults can occur as additive or multiplicative faults due to a malfunction or equipment aging.

For FDI, a distinction is usually made between additive and multiplicative faults. However, in FTC, the objective is to compensate for the fault effect on the system regardless of the nature of the fault.

The faults affecting a system are often represented by a variation of system parameters. Thus, in the presence of a fault, the system model can be written as

$$
\begin{cases}
x_f(k+1) = A_f x_f(k) + B_f u_f(k) \\
\qquad y_f(k) = C_f x_f(k)
\end{cases},
\tag{2.18}
$$

where the new matrices of the faulty system are defined by

$$
A_f = A + \delta A; \quad B_f = B + \delta B; \quad C_f = C + \delta C.
\tag{2.19}
$$

δA, δB, and δC correspond to the deviation of the system parameters with respect to the nominal values. However, when a fault occurs on the system, it is very difficult to get these new matrices on-line.

Process monitoring is necessary to ensure effectiveness of process control and consequently a safe and a profitable plant operation. As presented in the next paragraph, the effect of actuator and sensor faults can also be represented as an additional unknown input vector acting on the dynamics of the system or on the measurements.

The effect of actuator and sensor faults can also be represented using an unknown input vector $f_j \in \Re^l, j = a$ (for actuators), s (for sensors) acting on the dynamics of the system or on the measurements.

2.3.1 Actuator Faults

It is important to note that an actuator fault corresponds to the variation of the global control input U applied to the system, and not only to u:

$$U_f = \Gamma U + U_{f0}, \qquad (2.20)$$

where

- U is the global control input applied to the system
- U_f is the global faulty control input
- u is the variation of the control input around the operating point U_0, ($u = U - U_0$, $u_f = U_f - U_0$)
- U_{f0} corresponds to the effect of an additive actuator fault
- ΓU represents the effect of a multiplicative actuator fault

with $\Gamma = diag(\alpha)$, $\alpha = \begin{bmatrix} \alpha_1 & \cdots & \alpha_i & \cdots & \alpha_m \end{bmatrix}^T$ and $U_{f0} = \begin{bmatrix} u_{f01} & \cdots & u_{f0i} & \cdots & u_{f0m} \end{bmatrix}^T$. The ith actuator is faulty if $\alpha_i \neq 1$ or $u_{f0i} \neq 0$ as presented in Table 2.1 where different types of actuator faults are described.

Table 2.1. Actuator fault

	Constant offset $u_{f0i} = 0$	Constant offset $u_{f0i} \neq 0$
$\alpha_i = 1$	Fault-free case	Bias
$\alpha_i \in]0; 1[$	Loss of effectiveness	Loss of effectiveness
$\alpha_i = 0$	Out of order	Actuator blocked

In the presence of an actuator fault, the linearized system (2.10) can be given by

$$\begin{cases} x(k+1) = Ax(k) + B(\Gamma U(k) + U_{f0} - U_0) \\ \qquad y(k) = Cx(k) \end{cases}. \qquad (2.21)$$

The previous equation can also be written as

$$\begin{cases} x(k+1) = Ax(k) + Bu(k) + B[(\varGamma - I)U(k) + U_{f0}] \\ y(k) = Cx(k) \end{cases}. \tag{2.22}$$

By defining $f_a(k)$ as an unknown input vector corresponding to actuator faults, (2.18) can be represented as follows:

$$\begin{cases} x(k+1) = Ax(k) + Bu(k) + F_a f_a(k) \\ y(k) = Cx(k) \end{cases}, \tag{2.23}$$

where $F_a = B$ and $f_a = (\varGamma - I)U + U_{f0}$. If the ith actuator is declared to be faulty, then F_a corresponds to the ith column of matrix B and f_a corresponds to the magnitude of the fault affecting this actuator.

In the nonlinear case and in the presence of actuator faults, (2.3) can be described by the following continuous-time state-space representation:

$$\begin{cases} \dot{x}(t) = f(x(t)) + \sum_{j=1}^{m}(g_j(x(t))u_j(t)) + \sum_{j=1}^{m}(F_{a,j}(x(t))f_{a,j}(t)) \\ y_i(t) = h_i(x(t)) \qquad\qquad\qquad\qquad 1 \le i \le q \end{cases}, \tag{2.24}$$

where $F_{a,j}(x(t))$ corresponds to the jth column of matrix $G(x(t))$ in (2.3) and $f_{a,j}(t)$ corresponds to the magnitude of the fault affecting the jth actuator.

2.3.2 Sensor Faults

In a similar way, considering f_s as an unknown input illustrating the presence of a sensor fault, the linear faulty system will be represented by

$$\begin{cases} x(k+1) = Ax(k) + Bu(k) \\ y(k) = Cx(k) + F_s f_s(k) \end{cases}. \tag{2.25}$$

The affine nonlinear systems can be defined in continuous-time through an additive component such as

$$\begin{cases} \dot{x}(t) = f(x(t)) + G(x(t))u(t) \\ y_i(t) = h_i(x(t)) + F_{s,i}f_{s,i}(t) \end{cases}, \quad 1 \le i \le q, \tag{2.26}$$

where $F_{s,i}$ is the ith row of matrix F_s and $f_{s,i}$ is the fault magnitude affecting the ith sensor.

This description of actuator and sensor faults is a structured representation of these faults. Matrices F_a and F_s are assumed to be known and f_a and f_s correspond, respectively, to the magnitudes of the actuator fault and the sensor fault.

An FTC method is based on a nominal control law associated with a fault detection and estimation, and a modification of this control law. This is used in order to compensate for the fault effect on the system.

2.4 Nominal Tracking Control Law

The first step in designing an FTC method is the setup of a nominal control. In the sequel, a multi-variable linear tracking control is first addressed, then a case of nonlinear systems is presented.

2.4.1 Linear Case

The objective in this section is to describe a nominal tracking control law able to make the system outputs follow pre-defined reference inputs.

Consider a MIMO system given by the following discrete-time state-space representation:

$$\begin{cases} x(k+1) = Ax(k) + Bu(k) \\ \quad y(k) = Cx(k) \end{cases}, \tag{2.27}$$

where $x \in \Re^n$ is the state vector, $u \in \Re^m$ is the control input vector, and $y \in \Re^q$ is the output vector. A, B, and C are matrices of appropriate dimensions.

The tracking control law requires that the number of outputs to be controlled must be less than or equal to the number of the control inputs available on the system [29].

If the number of outputs is greater than the number of control inputs, the designer of the control law selects the outputs that must be tracked and breaks down the output vector y as follows:

$$y(k) = Cx(k) = \begin{bmatrix} C_1 \\ C_2 \end{bmatrix} x(k) = \begin{bmatrix} y_1(k) \\ y_2(k) \end{bmatrix}. \tag{2.28}$$

The feedback controller is required to cause the output vector $y_1 \in \Re^p$ ($p \leq m$) to track the reference input vector y_r such that in steady-state:

$$y_r(k) - y_1(k) = 0. \tag{2.29}$$

To achieve this objective, a comparator and integrator vector $z \in \Re^p$ is added to satisfy the following relation:

$$\begin{cases} z(k+1) = z(k) + T_s(y_r(k) - y_1(k)) \\ \qquad\quad = z(k) + T_s(y_r(k) - C_1 x(k)) \end{cases}, \tag{2.30}$$

where T_s is the sample period to be chosen properly. Careful consideration should be given to the choice of T_s. If T_s is too small, the processor will not have enough time to calculate the control law. The system may be unstable if T_s is too high because the system is operating in open-loop during a sample period.

The open-loop system is governed by the augmented state and output equations, where I_p is an identity matrix of dimension p and $0_{n,p}$ is a null matrix of n rows and p columns:

$$
\begin{cases}
\begin{bmatrix} x(k+1) \\ z(k+1) \end{bmatrix} = \begin{bmatrix} A & 0_{n,p} \\ -T_sC_1 & I_p \end{bmatrix} \begin{bmatrix} x(k) \\ z(k) \end{bmatrix} + \begin{bmatrix} B \\ 0_{p,m} \end{bmatrix} u(k) + \begin{bmatrix} 0_{n,p} \\ T_sI_p \end{bmatrix} y_r(k) \\
y(k) = \begin{bmatrix} C & 0_{q,p} \end{bmatrix} \begin{bmatrix} x(k) \\ z(k) \end{bmatrix}
\end{cases}
$$

$$(2.31)$$

This state-space representation can also be written in the following form:

$$
\begin{cases}
X(k+1) = \bar{A}X(k) + \bar{B}_uu(k) + \bar{B}_ry_r(k) \\
y(k) = \bar{C}X(k)
\end{cases}
$$

$$(2.32)$$

The nominal feedback control law of this system can be computed by

$$
u(k) = -KX(k) = -\begin{bmatrix} K_1 & K_2 \end{bmatrix} \begin{bmatrix} x(k) \\ z(k) \end{bmatrix}.
$$

$$(2.33)$$

$K = \begin{bmatrix} K_1 & K_2 \end{bmatrix}$ is the feedback gain matrix obtained, for instance, using a pole placement technique, linear-quadratic (LQ) optimization, and so on [6,78, 119,125,130,133]. To achieve this control law, the state variables are assumed to be available for measurement. Moreover, the state-space considered here is that where the outputs are the state variables (C is the identity matrix I_n). Otherwise, the control law is computed using the estimated state variables obtained, for instance, by an observer or a Kalman filter.

Figure 2.7 summarizes the design of the nominal tracking control taking into account the operating point with $x = y_1$.

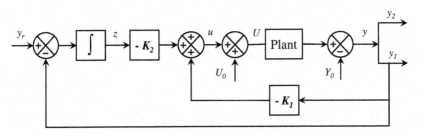

Fig. 2.7. Nominal tracking control taking into account the operating point

2.4.2 Nonlinear Case

The need for nonlinear control theory arises from the fact that systems are nonlinear in practice. Although linear models are simple and easy to analyze, they are not valid except around a certain operating region. Outside this region, the linear model is not valid and the linear representation of the process is insufficient. Control of nonlinear systems has been extensively considered in the literature where plenty of approaches have been proposed for deterministic, stochastic, and uncertain nonlinear systems (see for instance [46, 52, 58, 116]). In this book, two control methods are used: the exact input-output linearization and the sliding mode controller (SMC).

Exact Linearization and Decoupling Input-Output Controller

According to the special class of input-linear systems given by (2.3), a nonlinear control law is commonly established to operate in closed-loop. To perform this task, an exact linearization and decoupling input-output law via a static state-feedback $u(t) = \alpha(x(t)) + \beta(x(t)) v(t)$ is designed. It is assumed here that the system has as many outputs as inputs (*i.e.*, $q = m$). For the general case where $q \neq m$, the reader can refer to [41, 73, 98].

The aim of this control law is to transform (2.3) into a linear and controllable system based on the following definitions

Definition 2.1. *Let $(r_1, r_2, \ldots, r_m)$ be the set of the relative degree per row of (2.3) such as*

$$r_i = \{min\ l \in \aleph / \exists j \in [1, \ldots, m], L_{g_j} L_f^{l-1} h_i(x(t)) \neq 0\}, \tag{2.34}$$

where L is the Lie derivative operator.

The Lie derivative of h_i in the direction of f, denoted $L_f h_i(x)$, is the derivative of h_i in $t = 0$ along the integral curve of f, such that

$$L_f h_i(x) = \sum_{j=1}^{n} f_j(x) \frac{\partial h_i}{\partial x_j}(x). \tag{2.35}$$

The operation L_f, Lie derivative in the direction of f, can be iterated. $L_f^k h$ is defined for any $k \geq 0$ by

$$L_f^0 h(x) = h(x) \quad \text{and} \quad L_f^k h(x) = L_f(L_f^{k-1} h(x)) \quad \forall k \geq 1. \tag{2.36}$$

Definition 2.2. *If all r_i exist ($i = 1, \ldots, m$), the following matrix Δ is called "decoupling matrix" of (2.3):*

$$\Delta(x) = \begin{bmatrix} L_{g_1} L_f^{r_1-1} h_1(x) & \cdots & L_{g_m} L_f^{r_1-1} h_1(x) \\ \vdots & \ddots & \vdots \\ L_{g_1} L_f^{r_m-1} h_m(x) & \cdots & L_{g_m} L_f^{r_m-1} h_m(x) \end{bmatrix}. \tag{2.37}$$

A vector Δ_0 is also defined such as

$$\Delta_0(x) = \begin{bmatrix} L_f^{r_1} h_1(x) \\ \vdots \\ L_f^{r_m} h_m(x) \end{bmatrix}. \tag{2.38}$$

According to the previous definition, the nonlinear control is designed as follows.

Theorem:
a) The system defined by (2.3) is statically decouplable on a subset M_0 of $\Re^n$ if and only if

$$\text{rank } \Delta(x) = m, \qquad \forall x \in M_0. \tag{2.39}$$

b) The control law computed using the state-feedback is defined by

$$u(t) = \alpha(x(t)) + \beta(x(t))v(t), \tag{2.40}$$

where

$$\begin{cases} \alpha(x) = -\Delta^{-1}(x)\Delta_0(x) \\ \beta(x) = \Delta^{-1}(x) \end{cases}. \tag{2.41}$$

This control law is able to decouple (2.3) on M_0.
c) This closed-loop system has a linear input-output behavior described by

$$y_i^{(r_i)}(t) = v_i(t), \qquad \forall i \in [1, \ldots, m], \tag{2.42}$$

where $y_i^{(r_i)}(t)$ is the r_i^{th} derivative of y_i.

Two cases may be observed:

- $\sum_{i=1}^{m} r_i = n$: the closed-loop system characterized by the m decoupled linear subsystems is linear, controllable and observable.
- $\sum_{i=1}^{m} r_i < n$: a subspace made unobservable by the nonlinear feedback (2.40). The stability of the unobservable subspace must be studied. This subspace must have all modes stable. More details about this case can be found in [41, 73, 98].

Since each SISO linear subsystem is equivalent to a cascade of integrators, a second feedback control law should be considered in order to stabilize and to set the performance of the controlled nonlinear system. This second feedback is built using linear control theory [29]. The simplest feedback consists of using a pole placement associated with τ_i such as

$$\frac{y_i(s)}{y_{ref,i}(s)} = \frac{1}{(1 + \tau_i s)^{r_i}} \tag{2.43}$$

where $y_{ref,i}$ is the reference input associated with output y_i.

The advantage of this approach is that the feedback controllers are designed independently of each other. Indeed, nonlinear feedback (2.40) is built from model (2.3) according to the theorem stated previously. The stabilized feedback giving a closed-loop behavior described by (2.43) is designed from the m decoupled linear equivalent subsystems (2.42) written in the Brunovsky canonical form [65] such as

$$\begin{cases} \dot{z}_i(t) = A_i z_i(t) + B_i v_i(t) \\ y_i(t) = C_i z_i(t) \end{cases}, \qquad \forall i \in [1, \ldots, m], \tag{2.44}$$

with

$$A_i = \begin{bmatrix} 0 & & & \\ \vdots & & I_{r_i-1} & \\ 0 & & & \\ 0 & 0 & \cdots & 0 \end{bmatrix}, \quad B_i = \begin{bmatrix} 0 \\ \vdots \\ 0 \\ 1 \end{bmatrix}, \quad C_i = \begin{bmatrix} 1 & 0 & \cdots & 0 \end{bmatrix}. \tag{2.45}$$

The link between both state-feedbacks is defined by a diffeomorphism $z(t) = \Phi(x(t))$ where $z(t)$ is the state vector of the decoupled linear system written in the controllability canonical form.

When there is no unobservable state subspace, the diffeomorphism is defined by

$$z(t) = \Phi(x(t)) = \begin{bmatrix} \Phi_1(x(t)) \\ \vdots \\ \vdots \\ \vdots \\ \Phi_m(x(t)) \end{bmatrix} = \begin{bmatrix} \begin{bmatrix} h_1(x(t)) \\ \vdots \\ L_f^{r_1-1} h_1(x(t)) \end{bmatrix} \\ \vdots \\ \begin{bmatrix} h_m(x(t)) \\ \vdots \\ L_f^{r_m-1} h_m(x(t)) \end{bmatrix} \end{bmatrix}. \tag{2.46}$$

The exact linearization and decoupling input-output controller with both state-feedback control laws may be illustrated in Fig. 2.8.

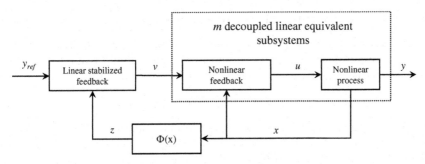

Fig. 2.8. Nonlinear control scheme

Sliding Mode Controller

The main advantage of the SMC over the other nonlinear control laws is its robustness to external disturbances, model uncertainties, and variations in system parameters [135, 136]. In order to explain the SMC concept, consider a SISO second order input affine nonlinear system:

$$\ddot{x} = f(x, \dot{x}) + g(x, \dot{x})u + d_f, \qquad (2.47)$$

where u is the control input and d_f represents the uncertainties and external disturbances which are assumed to be upper bounded with $|d_f| < D$. Note that this section considers continuous-time systems but the time index is omitted for simplicity. Defining the state variables as $x_1 = x$ and $x_2 = \dot{x}$, (2.47) leads to

$$\begin{cases} \dot{x}_1 = x_2 \\ \dot{x}_2 = f(x, \dot{x}) + g(x, \dot{x})u + d_f \end{cases}. \qquad (2.48)$$

If the desired trajectory is given as x_1^d, then the error between the actual x_1 and the desired trajectory x_1^d can be written as

$$e = x_1 - x_1^d. \qquad (2.49)$$

The time derivative of e is given by

$$\dot{e} = \dot{x}_1 - \dot{x}_1^d = x_2 - x_2^d. \qquad (2.50)$$

The switching surface s is conventionally defined for second order systems as a combination of the error variables e and $\dot{e}$:

$$s = \dot{e} + \lambda e, \qquad (2.51)$$

where λ sets the dynamics in the sliding phase ($s = 0$).

The control input u should be chosen so that trajectories approach the switching surface and then stay on it for all future time instants. Thus, the time derivative of s is given by

$$\dot{s} = f(x, \dot{x}) + g(x, \dot{x})u + d_f - \ddot{x}_1^{\,d} + \lambda\dot{e}. \tag{2.52}$$

The control input is expressed as the sum of two terms [114]. The first one, called the equivalent control, is chosen using the nominal plant parameters $(d_f = 0)$, so as to make $\dot{s} = 0$ when $s = 0$. It is given by [114]

$$u_{eq} = g(x, \dot{x})^{-1}(\ddot{x}_1^{\,d} - f(x, \dot{x}) - \lambda\dot{e}). \tag{2.53}$$

The second term is chosen to tackle the uncertainties in the system and to introduce a reaching law; the constant $(M sign(s))$ plus proportional (ks) rate reaching law is imposed by selecting the second term as [114]

$$u^* = g(x, \dot{x})^{-1}[-ks - M sign(s)], \tag{2.54}$$

where k and M are positive numbers to be selected and $sign(.)$ is the *signum* function. The function $g(x, \dot{x})$ must be invertible for (2.53) and (2.54) to hold.

Then, the control input $u = u_{eq} + u^*$ becomes

$$u = g(x, \dot{x})^{-1}[\ddot{x}_1^{\,d} - f(x, \dot{x}) - \lambda\dot{e} - ks - M sign(s)]. \tag{2.55}$$

Substituting input u of (2.55) in (2.52) gives the derivative $\dot{s}$ of the sliding surface

$$\dot{s} = -ks - M sign(s) + d_f. \tag{2.56}$$

The necessary condition for the existence of conventional sliding mode for (2.47) is given by

$$\frac{1}{2}\frac{d}{dt}s^2 < 0, \quad \text{or} \quad s\dot{s} < 0. \tag{2.57}$$

This condition states that the squared distance to the switching surface, as measured by s^2, decreases along all system trajectories. However, this condition is not feasible in practice because the switching of real components is not instantaneous and this leads to an undesired phenomenon known as chattering in the direction of the switching surface. Thus (2.57) is expanded by a boundary layer in which the controller switching is not required:

$$s\dot{s} < -\eta |s|. \tag{2.58}$$

Multiplying (2.56) by s yields

$$s\dot{s} = -ks^2 - M sign(s)s + d_f s = -ks^2 - M|s| + d_f s. \tag{2.59}$$

With a proper choice of k and M, (2.58) will be satisfied.

The elimination of the chattering effect produced by the discontinuous function *sign* is ensured by a saturation function *sat*. This saturation function is defined as follows:

$$sat(s) = \begin{cases} sign(s) & \text{if } |s| > \phi_s \\ s/\phi_s & \text{if } |s| < \phi_s \end{cases}, \tag{2.60}$$

where ϕ_s is a boundary layer around the sliding surface s.

2.5 Model-based Fault Diagnosis

After designing the nominal control law, it is important to monitor the behavior of the system in order to detect and isolate any malfunction as soon as possible. The FDI allows us to avoid critical consequences and helps in taking appropriate decisions either on shutting down the system safely or continuing the operation in degraded mode in spite of the presence of the fault.

The fault diagnostic problem from raw data trends is often difficult. However, model-based FDI techniques are considered and combined to supervise the process and to ensure appropriate reliability and safety in industry. The aim of a diagnostic procedure is to perform two main tasks: fault detection, which consists of deciding whether a fault has occurred or not, and fault isolation, which consists of deciding which element(s) of the system has (have) indeed failed. The general procedure comprises the following three steps:

- Residual generation: the process of associating, with the measured and estimated output pair $(y, \widehat{y})$, features that allow the evaluation of the difference, denoted r $(r = y - \widehat{y})$, with respect to normal operating conditions
- Residual evaluation: the process of comparing residuals r to some predefined thresholds according to a test and at a stage where symptoms $S(r)$ are produced
- Decision making: the process of deciding through an indicator, denoted I based on the symptoms $S(r)$, which elements are faulty (*i.e.*, isolation)

This implies the design of residuals r that are close to zero in the fault-free situations $(f = 0)$, while they will clearly deviate from zero in the presence of faults $(f \neq 0)$. They will possess the ability to discriminate between all possible modes of faults, which explains the use of the term isolation. A short historical review of FDI can also be found in [71] and current developments are reviewed in [44].

While a single residual may be enough to detect a fault, a set of structured residuals is required for fault isolation . In order to isolate a fault, some residuals with particular sensitivity properties are established. This means that $r = 0$ if $f^* = 0$ and $r \neq 0$ if $f^* \neq 0$ regardless of the other faults defined through $f^d = 0$. In this context, in order to isolate and to estimate both actuator and sensor faults, a bank of structured residuals is considered where each residual vector r may be used to detect a fault according to a statistical test. Consequently, it involves the use of statistical tests such as the Page-Hinkley test, limit checking test, generalized likelihood ratio test, and trend analysis test [8].

An output vector of the statistical test, called *coherence vector* S_r, can then be built from the bank of ν residual generators:

$$S_r = [S(||r_1||) \cdots S(||r_\nu||)]^T, \tag{2.61}$$

where $S(\|r_j\|)$ represents a symptom associated with the norm of the residual vector r_j. It is equal to 0 in the fault-free case and set to 1 when a fault is detected.

The coherence vector is then compared to the fault signature vector S_{ref,f_j} associated with the jth fault according to the residual generators built to produce a signal sensitive to all faults except one as represented in Table 2.2.

Table 2.2. Fault signature table

S_r	No faults	S_{ref,f_1}	S_{ref,f_2}	$\cdots$	S_{ref,f_ν}	Other faults
$S(\|r_1\|)$	0	0	1	$\cdots$	1	1
$S(\|r_2\|)$	0	1	0	$\cdots$	1	1
$\vdots$	$\vdots$	$\vdots$	$\vdots$	$\ddots$	$\vdots$	$\vdots$
$S(\|r_\nu\|)$	0	1	1	$\cdots$	0	1

The decision is then made according to an elementary logic test [86] that can be described as follows: an indicator $I(f_j)$ is equal to 1 if S_r is equal to the jth column of the incidence matrix (S_{ref,f_j}) and otherwise it is equal to 0. The element associated with the indicator equals to 1 is then declared to be faulty.

Moreover, the FDI module can also be exploited in order to estimate the fault magnitude.

Based on a large diversity of advanced model-based methods for automated FDI [22,31,53,69], the problem of actuator and/or sensor fault detection and magnitude estimation for both linear time-invariant (LTI) and nonlinear systems has been considered in the last few decades. Indeed, due to difficulties inherent in the on-line identification of closed-loop systems, parameter estimation techniques are not considered in this book. The parity space technique is suitable to distinguish between different faults in the presence of uncertain parameters, but is not useful for fault magnitude estimation.

In this section, the FDI problem is first considered, then in Sect. 2.6 the fault estimation is treated before investigating the FTC problem in Sect. 2.7.

2.5.1 Actuator/Sensor Fault Representation

Let us recall the state-space representation of a system that may be affected by actuator and/or sensor fault:

$$\begin{cases} x(k+1) = Ax(k) + Bu(k) + F_a f_a(k) \\ \quad\;\; y(k) = Cx(k) + F_s f_s(k) \end{cases}, \qquad (2.62)$$

where matrices F_a and F_s are assumed to be known and f_a and f_s correspond to the magnitude of the actuator and the sensor faults, respectively. The magnitude and time occurrence of the faults are assumed to be completely unknown.

In the presence of sensor and actuator faults, (2.62) can also be represented by the unified general formulation

$$\begin{cases} x(k+1) = Ax(k) + Bu(k) + F_x f(k) \\ \quad\quad y(k) = Cx(k) + F_y f(k) \end{cases}, \tag{2.63}$$

where $f = [f_a^T \ f_s^T]^T \in \Re^\nu$ ($\nu = m + q$) is a common representation of sensor and actuator faults. $F_x \in \Re^{n\times\nu}$ and $F_y \in \Re^{q\times\nu}$ are respectively the actuator and sensor faults matrices with $F_x = [B \ 0_{n\times q}]$ and $F_y = [0_{q\times m} \ I_q]$.

The objective is to isolate faults. This is achieved by generating residuals sensitive to certain faults and insensitive to others, commonly called structured residuals . The fault vector f in (2.63) can be split into two parts. The first part contains the "d" faults to be isolated $f^0 \in \Re^d$. In the second part, the other "$\nu - d$" faults are gathered in a vector $f^* \in \Re^{\nu-d}$. Then, the system can be written by the following equations:

$$\begin{cases} x(k+1) = Ax(k) + Bu(k) + F_x^0 f^0(k) + F_x^* f^*(k) \\ \quad\quad y(k) = Cx(k) + F_y^0 f^0(k) + F_y^* f^*(k) \end{cases}. \tag{2.64}$$

Matrices F_x^0, F_x^*, F_y^0, and F_y^*, assumed to be known, characterize the distribution matrices of f^* and f^0 acting directly on the system dynamics and on the measurements, respectively.

As indicated previously, an FDI procedure is developed to enable the detection and the isolation of a particular fault f^0 among several others. In order to build a set of residuals required for fault isolation, a residual generation using an unknown input decoupled scheme is considered such that the residuals are sensitive to fault vector f^* and insensitive to f^0. Only a single fault (actuator or sensor fault) is assumed to occur at a given time, because simultaneous faults can hardly be isolated. Hence, vector f^0 is a scalar ($d = 1$) and it is considered as an unknown input. It should be noted that the necessary condition of the existence of decoupled residual generator is fulfilled according to Hou and Muller [66]: the number of unknown inputs must be less than the number of measurements ($d \leq q$).

In case of an ith actuator fault, the system can be represented according to (2.64) by

$$\begin{cases} x(k+1) = Ax(k) + Bu(k) + B_i f^0(k) + [\overline{B}_i \ 0_{n\times q}] f^*(k) \\ \quad\quad y(k) = Cx(k) + [0_{q\times(p-1)} \ I_q] f^*(k) \end{cases}, \tag{2.65}$$

where B_i is the ith column of matrix B and $\overline{B}_i$ is matrix B without the ith column.

In order to generate a unique representation, (2.65) can be described as:

$$\begin{cases} x(k+1) = Ax(k) + Bu(k) + F_d f_d(k) + F_x^* f^*(k) \\ \quad\quad y(k) = Cx(k) + F_y^* f^*(k) \end{cases}, \quad (2.66)$$

where f^0 is denoted as f_d.

Similarly, for a jth sensor fault, the system is described as follows:

$$\begin{cases} x(k+1) = Ax(k) + Bu(k) + [B\ \ 0_{n\times(q-1)}]f^*(k) \\ \quad\quad y(k) = Cx(k) + E_j f^0(k) + [0_{q\times p}\ \ \overline{E}_j]f^*(k) \end{cases}, \quad (2.67)$$

where $E_j = [0 \cdots 1 \cdots 0]^T$ represents the jth sensor fault effect on the output vector and $\overline{E}_j$ is the identity matrix without the jth column.

According to Park *et al.* [100], a system affected by a sensor fault can be written as a system represented by an actuator fault. Assume the dynamic of a sensor fault is described as

$$f^0(k+1) = f^0(k) + T_s \xi(k), \quad (2.68)$$

where ξ defines the sensor error input and T_s is the sampling period.

From (2.67) and (2.68), a new system representation including the auxiliary state can be introduced:

$$\begin{cases} \begin{bmatrix} x(k+1) \\ f^0(k+1) \end{bmatrix} = \begin{bmatrix} A & 0_{n\times 1} \\ 0_{1\times n} & 1 \end{bmatrix} \begin{bmatrix} x(k) \\ f^0(k) \end{bmatrix} + \begin{bmatrix} B \\ 0_{1\times m} \end{bmatrix} u(k) + \begin{bmatrix} 0_{n\times 1} \\ T_s \end{bmatrix} \xi(k) \\ \quad\quad + \begin{bmatrix} B & 0_{n\times(q-1)} \\ 0_{1\times m} & 0_{1\times(q-1)} \end{bmatrix} f^*(k) \\ y(k) = \begin{bmatrix} C & E_j \end{bmatrix} \begin{bmatrix} x(k) \\ f^0(k) \end{bmatrix} + \begin{bmatrix} 0_{q\times m} & \overline{E}_j \end{bmatrix} f^*(k) \end{cases}.$$

$$(2.69)$$

Consequently, for actuator or sensor faults representation ((2.65) and (2.69)), a unique state-space representation can be established to describe the faulty system as follows:

$$\begin{cases} x(k+1) = Ax(k) + Bu(k) + F_d f_d(k) + F_x^* f^*(k) \\ \quad\quad y(k) = Cx(k) + F_y^* f^*(k) \end{cases}, \quad (2.70)$$

where f_d is the unknown input vector. For simplicity, the same notation for vectors and matrices has been used in (2.66) and (2.70).

Under the FTC framework, once the FDI module indicates which sensor or actuator is faulty, the fault magnitude should be estimated and a new control law will be set up in order to compensate for the fault effect on the system.

As sensor and actuator faults have different effects on the system, the control law should be modified according to the nature of the fault. In this book, only one fault is assumed to occur at a given time. The presence of simultaneous multiple faults is still rare, and the FDI problem in this case is considered as a specific topic and is dealt with in the literature. Here, the objective is to deal with a complete FTC problem for a single fault.

2.5.2 Residual Generation

Unknown Input Observer – Linear Case

Based on the previous representation, several approaches have been suggested by [43, 53] to generate a set of residuals called structural residuals, in order to detect and isolate the faulty components. The theory and the design of unknown input observers developed in [22] is considered in this book due to the fact that a fault magnitude estimation can be generated but also that the unknown observers concept can be extended to nonlinear systems. A full-order observer is built as follows:

$$\begin{cases} w(k+1) = Ew(k) + TBu(k) + Ky(k) \\ \widehat{x}(k) = w(k) + Hy(k) \end{cases}, \qquad (2.71)$$

where $\widehat{x}$ is the estimated state vector and w is the state of this full-order observer. E, T, K, and H are matrices to be designed for achieving unknown input decoupling requirements. The state estimation error vector $(e = \widehat{x} - x)$ of the observer goes to zero asymptotically, regardless of the presence of the unknown input in the system. The design of the unknown input observer is achieved by solving the following equations:

$$(HC - I)F_d = 0, \qquad (2.72)$$

$$T = I - HC, \qquad (2.73)$$

$$E = A - HCA - K_1C, \qquad (2.74)$$

$$K_2 = EH, \qquad (2.75)$$

and

$$K = K_1 + K_2. \qquad (2.76)$$

E must be a stable matrix in order to guarantee a state error estimation equal to zero.

The system defined by (2.71) is an unknown input observer for the system given by (2.70) if the necessary and sufficient conditions are fulfilled:

- Rank(CF_d) = rank(F_d)
- (C, A_1) is a detectable pair, where $A_1 = E + K_1C$

If these conditions are fulfilled, an unknown input observer provides an estimation of the state vector, used to generate a residual vector $r(k) = y(k) - C\widehat{x}(k)$ independent of $f_d(k)$. This means that $r(k) = 0$ if $f^*(k) = 0$ and $r(k) \neq 0$ if $f^*(k) \neq 0$ for all $u(k)$ and $f_d(k)$.

Unknown Input Observer – Affine Case

Among all algebraic methods, several methods consist of the generation of fault decoupling residual for special class of nonlinear systems such as bilinear systems [80]. Other methods focus more on general nonlinear systems where an unknown input decoupling input-output model is obtained [138]. Exact fault decoupling for nonlinear systems is also synthesized with geometric approach by [60, 101]. A literature review is detailed in [81].

Consider the state-space representation of the affine system affected by an actuator fault:

$$\begin{cases} \dot{x}(t) = f(x(t)) + \sum_{j=1}^{m} g_j(x(t))u_j(t) + \sum_{j=1}^{m} F_j(x(t))f_j(t) \\ y_i(t) = h_i(x(t)), \qquad\qquad 1 \le i \le q \end{cases} . \tag{2.77}$$

The approach presented in this section is an extension of the synthesis of unknown input linear observers to affine nonlinear systems. The initial work on this problem can be found in [49, 50].

The original system described by (2.77) should be broken down into two subsystems where one subsystem depends on the fault vector f and the second is independent of f by means of a diffeomorphism Φ_f such as

$$\begin{cases} \dot{\tilde{x}}_1(t) = \tilde{f}_1(\tilde{x}_1(t), \tilde{x}_2(t)) + \sum_{j=1}^{m} \tilde{g}_{1j}(\tilde{x}_1(t), \tilde{x}_2(t))u_j(t) \\ \qquad\qquad\qquad + \sum_{j=1}^{m} \tilde{F}_j(\tilde{x}_1(t), \tilde{x}_2(t))f_j(t) , \\ \dot{\tilde{x}}_2(t) = \tilde{f}_2(\tilde{x}_1(t), \tilde{x}_2(t)) + \sum_{j=1}^{m} \tilde{g}_{2j}(\tilde{x}_1(t), \tilde{x}_2(t))u_j(t) \end{cases} \tag{2.78}$$

where $\tilde{x}(t) = \begin{bmatrix} \tilde{x}_1(t) \\ \tilde{x}_2(t) \end{bmatrix} = \Phi_f(x(t), u(t))$.

The diffeomorphism Φ_f is defined by

$$\sum_{j=1}^{m} \frac{\partial}{\partial x_j(t)} \Phi_f(x(t), u(t)) \times F_j(x(t))f_j(t) = 0 . \tag{2.79}$$

This transformation is solved using the Frobenius theorem [73]. Equation (2.79) is not always satisfied. In order to simplify the way to solve this transformation, only one component $f_j(t)$ of the fault vector f is considered with the objective of building a bank of observers. Each observer is dedicated to one single fault f_j as proposed in the generalized observer scheme.

A subsystem insensitive to a component f_j of the fault vector $f(t)$ is extracted for each observer by deriving the output vector $y(t)$. A characteristic index is associated with each fault f_j. This index corresponds to the necessary derivative number so that the fault f_j appears in y_i. This index is also called *the detectability index* and is defined by

$$\rho_i = \min\{\zeta \in \mathbb{N} | L_F L_g^{\zeta-1} h_i(x(t)) \neq 0\}. \tag{2.80}$$

If ρ_i exists, only component output y_i is affected by f_j. It is then possible to define a new state-space representation where a subsystem is insensitive to fault f_j, such as

$$\widetilde{x}(t) = \Phi_{f_j}(x(t), u(t)) = \begin{bmatrix} \widetilde{x}_1(t) \\ \widetilde{x}_2(t) \end{bmatrix} = \begin{bmatrix} \begin{bmatrix} y_i(t) \\ \dot{y}_i(t) \\ \vdots \\ y_i^{\rho_i-1}(t) \\ \phi_i(x(t), u(t)) \end{bmatrix} \end{bmatrix}. \tag{2.81}$$

It is always possible to find $\phi_i(x(t), u(t))$ satisfying the following conditions [41]:

$$\text{rank}\left(\frac{\partial}{\partial x} \left[\begin{bmatrix} y_i(t) \\ \dot{y}_i(t) \\ \vdots \\ y_i^{\rho_i-1}(t) \\ \phi_i(x(t), u(t)) \end{bmatrix} \right] \right) = \dim(x(t)), \tag{2.82}$$

where $\frac{d}{dt}(\phi_i(x(t), u(t)))$ is independent of $f_j(t)$.

System (2.77) can now be written by means of the new coordinates system defined in (2.81) and a subsystem insensitive to f_j can be represented as

$$\begin{cases} \dot{\widetilde{x}}_1(t) = \phi_i(\widetilde{x}_1(t), \widetilde{x}_2(t), u(t)) \\ \widetilde{y}_i(t) = \widetilde{h}_i(\widetilde{x}_1(t), \widetilde{x}_2(t)) \end{cases}, \tag{2.83}$$

where $\widetilde{y}_i(t)$ is the output vector $y(t)$ without the ith component $y_i(t)$. $\widetilde{x}_2(t)$ is considered as an input vector for (2.83).

Considering all the components of the fault vector $f(t)$, a bank of observers is built where each observer is insensitive to a unique fault f_j. Nonlinear subsystem (2.83), which is insensitive to f_j, is used in order to synthesize a nonlinear observer as an extended Luenberger observer [97].

Based on [100], the proposed decoupled observer method applied to an affine system also provides an efficient FDI technique for sensor faults as developed in the linear case. In the presence of a sensor fault, the observer insensitive to the fault estimates state vector $\widetilde{x}_1(t)$ and consequently estimates the output corrupted by the fault. On the other hand, no estimation of an actuator fault can be computed from (2.83).

Fault Diagnosis Filter Design

Some control methods such as observers have been considered or modified to solve FDI problems. Among various FDI methods, filters have been successfully considered to provide new tools to detect and isolate faults.

To detect and estimate the fault magnitude, a fault detection filter is designed such that it does not decouple the residuals from the fault but rather assigns the residuals vector in particular directions to guarantee the identification of the fault [28, 79, 122].

Under the condition that (A, C) is observable from (2.66) or (2.70), the projectors are designed such that the residual vector is sensitive only to a particular fault direction. In order to determine the fault magnitude and the state vector estimations, a gain is synthesized such that the residual vector $r(k) = y(k) - C\widehat{x}(k)$ is insensitive to specific faults according to some projectors P. These projectors are designed such that the projected residual vector $p(k) = Pr(k)$ is sensitive only to a particular fault direction. Hence, the specific fault filter is defined as follows:

$$\begin{cases} \widehat{x}(k+1) = A\widehat{x}(k) + Bu(k) + (K_A + K_C)(y(k) - C\widehat{x}(k)) \\ \widehat{y}(k) = C\widehat{x}(k) \end{cases}, \qquad (2.84)$$

where

- K_A should be defined in order to obtain $AF_d - K_A CF_d = 0$, so K_A is equivalent to

$$K_A = \omega \Xi, \qquad (2.85)$$

  $\omega = AF_d$, $\Xi = (CF_d)^+$ and $+$ defines the pseudo-inverse
- K_C should be defined in order to obtain $K_C CF_d = 0$ which is solved as follows:

$$K_C = K\Psi, \qquad (2.86)$$

  where $\Psi = \beta \left[I - (CF_d)(CF_d)^+ \right]$ and K is a constant gain

It must be noted that β is chosen as a matrix with appropriate dimensions whose elements are equal to 1. The reduced gain K defines the unique free parameter in this specific filter.

Based on (2.85) and (2.86), (2.84) becomes equivalent to the following:

$$\begin{cases} \widehat{x}(k+1) = (\mathcal{A} - K\mathcal{C})\widehat{x}(k) + Bu(k) + K_A y_k + K\Psi y_k \\ \widehat{y}(k) = C\widehat{x}(k) \end{cases}, \qquad (2.87)$$

where $\mathcal{A} = A[I - F_d \Xi C]$ and $\mathcal{C} = \Psi C$.

The gain K is calculated using the eigenstructure assignment method such that $(\mathcal{A} - K\mathcal{C})$ is stable.

The gain breakdown $K_A + K_C$ and associated definitions involve the following matrices properties:

$$\Xi CF_d = 0, \qquad \text{and} \qquad \Psi CF_d = I, \qquad (2.88)$$

and enable the generation of projected residual vector as follows:

$$p(k) = Pr(k) = \begin{bmatrix} \Psi \\ \Xi \end{bmatrix} r(k) = \begin{bmatrix} \Sigma r(k) \\ \Xi r(k) + f_d(k-1) \end{bmatrix} = \begin{bmatrix} \gamma(k) \\ \eta(k) \end{bmatrix}. \qquad (2.89)$$

It is worth noting that γ is a residual insensitive to faults and η is calculated in order to be sensitive to f_d.

As sensor and actuator faults do not affect the system similarly, the control law should be modified according to the nature of the fault. In the sequel, different methods for estimating the actuator and sensor faults are presented.

2.6 Actuator and Sensor Faults Estimation

2.6.1 Fault Estimation Based on Unknown Input Observer

According to the fault isolation, the fault magnitude estimation of the corrupted element is extracted directly from the jth unknown input observer which is built to be insensitive to the jth fault ($f^*(k) = 0$). Based on the unknown input observer, the substitution of the state estimation in the faulty description (2.70) leads to

$$F_d f_d(k) = \hat{x}(k+1) - A\hat{x}(k) - Bu(k). \qquad (2.90)$$

In the presence of an actuator fault, F_d is a matrix of full column rank. Thus, the estimation of the fault magnitude $\hat{f}_0(k) = \hat{f}_d(k)$ makes use of the singular-value decomposition (SVD) [54].

Let $F_d = U \begin{bmatrix} R \\ 0 \end{bmatrix} V^T$ be the SVD of F_d. Thus, R is a diagonal and nonsingular matrix and U and V are orthonormal matrices.

Using the SVD and substituting it in (2.90) results in

$$\overline{\overline{x}}(k+1) = \overline{A\hat{x}}(k) + \overline{B\hat{u}}(k) + \begin{bmatrix} R \\ 0 \end{bmatrix} V^T f_d(k), \qquad (2.91)$$

where

$$\hat{x}(k) = U\overline{\overline{x}}(k) = U \begin{bmatrix} \overline{\overline{x}}_1(k) \\ \overline{\overline{x}}_2(k) \end{bmatrix}, \qquad (2.92)$$

$$\overline{A} = U^{-1}AU = \begin{bmatrix} \overline{A}_{11}(k) & \overline{A}_{12}(k) \\ \overline{A}_{21}(k) & \overline{A}_{22}(k) \end{bmatrix}, \qquad (2.93)$$

and

$$\overline{B} = U^{-1}B = \begin{bmatrix} \overline{B}_1(k) \\ \overline{B}_2(k) \end{bmatrix}. \tag{2.94}$$

Based on (2.91), the estimation of the actuator fault magnitude is defined as

$$\widehat{f}^0(k) = \widehat{f}_d(k) = VR^{-1}(\widehat{\overline{x}}_1(k+1) - \overline{A}_{11}\widehat{\overline{x}}_1(k) - \overline{A}_{12}\widehat{\overline{x}}_2(k) - \overline{B}_1 u(k)). \tag{2.95}$$

For a sensor fault, the fault estimation $\widehat{f}^0(k)$ is the last component of the estimated augmented state vector $\widehat{x}(k)$ as defined in (2.69).

2.6.2 Fault Estimation Based on Decoupled Filter

Based on the projected residual $p(k)$, an estimation of input vector $\eta(k)$ (which corresponds to the fault magnitude with a delay of one sample) should be directly exploited for fault detection. Indeed, a residual evaluation algorithm can be performed by the direct fault magnitude evaluation through a statistical test in order to monitor the process. It should be highlighted that the first component of projector vector (2.89), denoted $\gamma(k)$, can be considered as a quality indicator of the FDI module. If a fault is not equal to f_d then the mean of the indicator will not equal zero. As previously, sensor fault estimation can be also provided by the last component of the augmented state-space .

2.6.3 Fault Estimation Using Singular Value Decomposition

Another method to estimate the actuator and sensor faults is based on SVD which will be described in this section.

Estimation of Actuator Faults

In the presence of an actuator fault and according to (2.23) and (2.31), the augmented state-space representation of the system is written as

$$\begin{cases} \begin{bmatrix} x(k+1) \\ z(k+1) \end{bmatrix} = \begin{bmatrix} A & 0_{n,p} \\ -T_s C_1 & I_p \end{bmatrix} \begin{bmatrix} x(k) \\ z(k) \end{bmatrix} + \begin{bmatrix} B \\ 0_{p,m} \end{bmatrix} u(k) \\ \qquad\qquad + \begin{bmatrix} 0_{n,p} \\ T_s I_p \end{bmatrix} y_r(k) + \begin{bmatrix} F_a \\ 0 \end{bmatrix} f_a(k) \ , \\ y(k) = \begin{bmatrix} C & 0_{q,p} \end{bmatrix} \begin{bmatrix} x(k) \\ z(k) \end{bmatrix} \end{cases} \tag{2.96}$$

where F_a corresponds to the ith column of matrix B in case the ith actuator is faulty.

The magnitude of the fault f_a can be estimated if it is defined as a component of an augmented state vector $\overline{X}_a(k)$. In this case, the system (2.96) can be re-written under the following form:

$$\overline{E}_a\overline{X}_a(k+1) = \overline{A}_a\overline{X}_a(k) + \overline{B}_a\overline{U}(k) + \overline{G}_ay_r(k), \qquad (2.97)$$

where

$$\overline{E}_a = \begin{bmatrix} I_n & 0 & -F_a \\ 0 & I_p & 0 \\ C & 0 & 0 \end{bmatrix}; \quad \overline{A}_a = \begin{bmatrix} A & 0 & 0 \\ -T_sC_1 & I_p & 0 \\ 0 & 0 & 0 \end{bmatrix}; \quad \overline{B}_a = \begin{bmatrix} B & 0 \\ 0 & 0 \\ 0 & I_q \end{bmatrix};$$

$$\overline{G}_a = \begin{bmatrix} 0 \\ T_sI_p \\ 0 \end{bmatrix}; \quad \overline{X}_a(k) = \begin{bmatrix} x(k) \\ z(k) \\ f_a(k-1) \end{bmatrix}; \quad \overline{U}(k) = \begin{bmatrix} u(k) \\ y(k+1) \end{bmatrix}.$$

The estimation of the fault magnitude f_a can then be obtained using the SVD of matrix $\overline{E}_a$ if it is of full column rank [9].

Consider the SVD of matrix $\overline{E}_a$:

$$\overline{E}_a = T\begin{bmatrix} S \\ 0 \end{bmatrix}M^T, \quad \text{with} \quad T = \begin{bmatrix} T_1 & T_2 \end{bmatrix}.$$

T and M are orthonormal matrices such that: $TT^T = I$, $MM^T = I$, and S is a diagonal nonsingular matrix.

Substituting the SVD of $\overline{E}_a$ in (2.97) leads to

$$\begin{cases} \overline{X}_a(k+1) = \tilde{A}_a\overline{X}_a(k) + \tilde{B}_a\overline{U}(k) + \tilde{G}_ay_r(k) \\ 0 = \tilde{A}_0\overline{X}_a(k) + \tilde{B}_0\overline{U}(k) + \tilde{G}_0y_r(k) \end{cases}, \qquad (2.98)$$

where

$$\begin{aligned} \tilde{A}_a &= MS^{-1}T_1^T\,\overline{A}_a = \overline{E}_a^+\overline{A}_a; & \tilde{A}_0 &= T_2^T\overline{A}_a; \\ \tilde{B}_a &= MS^{-1}T_1^T\,\overline{B}_a = \overline{E}_a^+\overline{B}_a; & \tilde{B}_0 &= T_2^T\overline{B}_a; \qquad (2.99) \\ \tilde{G}_a &= MS^{-1}T_1^T\,\overline{G}_a = \overline{E}_a^+\overline{G}_a; & \tilde{G}_0 &= T_2^T\overline{G}_a; \end{aligned}$$

and where $\overline{E}_a^+$ is the pseudo-inverse of matrix $\overline{E}_a$.

Therefore, the estimation $\hat{f}_a$ of the fault magnitude f_a is the last component of the state vector $\overline{X}_a$, which is the solution of the first equation in (2.98). This solution $\overline{X}_a$ must satisfy the second equation of (2.98). It can be noted from (2.97) that the estimation of the fault magnitude f_a at time instant (k) depends on the system outputs y at time instant $(k+1)$. To avoid this problem, the computation of the fault estimation is delayed by one sample.

Estimation of Sensor Faults

When a sensor fault affects the closed-loop system, the tracking error between the reference input and the measurement will no longer be equal to zero. In this case, the nominal control law tries to bring the steady-state error back to zero. Hence, in the presence of a sensor fault, the control law must be prevented from reacting, unlike the case of an actuator fault. This can be achieved by cancelling the fault effect on the control input.

For sensor faults, the output equation given in (2.25) is broken down according to (2.28), and can be written as

$$y(k) = Cx(k) + F_s f_s(k) = \begin{bmatrix} y_1(k) \\ y_2(k) \end{bmatrix} = \begin{bmatrix} C_1 \\ C_2 \end{bmatrix} x(k) + \begin{bmatrix} F_{s1} \\ F_{s2} \end{bmatrix} f_s(k). \quad (2.100)$$

In this case, attention should be paid to the integral error vector z which will be affected by the fault as well. The integral error vector can then be described as follows:

$$\begin{cases} z(k+1) = z(k) + T_s(y_r(k) - y_1(k)) \\ \qquad = z(k) + T_s(y_r(k) - C_1 x(k) - F_{s1} f_s(k)) \end{cases}. \quad (2.101)$$

The sensor fault magnitude can be estimated in a similar way to that of the actuator fault estimation by describing the augmented system as follows:

$$\overline{E}_s \overline{X}_s(k+1) = \overline{A}_s \overline{X}_s(k) + \overline{B}_s \overline{U}(k) + \overline{G}_s y_r(k), \quad (2.102)$$

where

$$\overline{E}_s = \begin{bmatrix} I_n & 0 & 0 \\ 0 & I_p & 0 \\ C & 0 & F_s \end{bmatrix}; \quad \overline{A}_s = \begin{bmatrix} A & 0 & 0 \\ -T_s C_1 & I_p & -T_s F_{s1} \\ 0 & 0 & 0 \end{bmatrix}; \quad \overline{B}_s = \begin{bmatrix} B & 0 \\ 0 & 0 \\ 0 & I_q \end{bmatrix};$$

$$\overline{G}_s = \begin{bmatrix} 0 \\ T_s I_p \\ 0 \end{bmatrix}; \quad \overline{X}_s(k) = \begin{bmatrix} x(k) \\ z(k) \\ f_s(k) \end{bmatrix}; \quad \overline{U}(k) = \begin{bmatrix} u(k) \\ y(k+1) \end{bmatrix}.$$

The sensor fault magnitude $\hat{f}_s$ can then be estimated using the SVD of matrix $\overline{E}_s$ if this matrix is of full column rank.

2.7 Actuator and Sensor Fault-tolerance Principles

2.7.1 Compensation for Actuator Faults

The effect of the actuator fault on the closed-loop system is illustrated by substituting the feedback control law (2.33) in (2.23):

$$\begin{cases} x(k+1) = (A - BK_1)x(k) - BK_2z(k) + F_af_a(k) \\ \qquad y(k) = Cx(k) \end{cases}. \qquad (2.103)$$

A new control law u_{add} should be calculated and added to the nominal one in order to compensate for the fault effect on the system. Therefore, the total control law applied to the system is given by

$$u(k) = -K_1x(k) - K_2z(k) + u_{add}(k). \qquad (2.104)$$

Considering this new control law given by (2.104), the closed-loop state equation becomes

$$x(k+1) = (A - BK_1)x(k) - BK_2z(k) + F_af_a(k) + Bu_{add}(k). \qquad (2.105)$$

From this last equation, the additive control law u_{add} must be computed such that the faulty system is as close as possible to the nominal one. In other words, u_{add} must satisfy

$$Bu_{add}(k) + F_af_a(k) = 0. \qquad (2.106)$$

Using the estimation of the fault magnitude described in the previous section, the solution of (2.106) can be obtained by the following relation if matrix B is of full row rank:

$$u_{add}(k) = -B^{-1}F_a\hat{f}_a(k). \qquad (2.107)$$

The fault compensation principle presented under linear assumption can be directly extended to nonlinear affine systems but not to general ones. Indeed, according to (2.42) an additional control law can be applied to the decoupled linear subsystems. The three-tank system considered in Chap. 4 will provide an excellent example to illustrate this FTC design.

Remark 2.1. Matrix B is of full row rank if the number of control inputs is equal to the number of state variables. In this case, B is invertible.

Case of Non Full Row Rank Matrix B

In the case when matrix B is not of full row rank (*i.e.*, the number of system inputs is less than the number of system states), the designer chooses to maintain as many priority outputs as available control inputs to the detriment of other secondary outputs. To be as close as possible to the original system, these priority outputs are composed of the tracked outputs and of other remaining outputs. This is achieved at the control law design stage using, if necessary, a transformation matrix P such that

$$\begin{cases} \begin{bmatrix} x_p(k+1) \\ x_s(k+1) \end{bmatrix} = \begin{bmatrix} A_{pp} & A_{ps} \\ A_{sp} & A_{ss} \end{bmatrix} \begin{bmatrix} x_p(k) \\ x_s(k) \end{bmatrix} + \begin{bmatrix} B_p \\ B_s \end{bmatrix} u(k) + \begin{bmatrix} F_{ap} \\ F_{as} \end{bmatrix} f_a(k) \\ \quad y(k) = \begin{bmatrix} y_p(k) \\ y_s(k) \end{bmatrix} = C_T \begin{bmatrix} x_p(k) \\ x_s(k) \end{bmatrix} \end{cases},$$

$$(2.108)$$

where index p represents the priority variables and s corresponds to the secondary variables. In this way, B_p is a nonsingular square matrix. If the state-feedback gain matrix K_1 is broken down into $K_1 = \begin{bmatrix} K_p & K_s \end{bmatrix}$, the control law is then given by

$$u(k) = - \begin{bmatrix} K_p & K_s & K_2 \end{bmatrix} \begin{bmatrix} x_p(k) \\ x_s(k) \\ z(k) \end{bmatrix} + u_{add}(k). \qquad (2.109)$$

Substituting (2.109) in (2.108) leads to

$$\begin{cases} x_p(k+1) = (A_{pp} - B_p K_p)x_p(k) - B_p K_2 z(k) \\ \qquad + (A_{ps} - B_p K_s)x_s(k) + F_{ap}f_a(k) + B_p u_{add}(k) \end{cases} \qquad (2.110)$$

and

$$\begin{cases} x_s(k+1) = (A_{ss} - B_s K_s)x_s(k) - B_s K_2 z(k) \\ \qquad + (A_{sp} - B_s K_p)x_p(k) + F_{as}f_a(k) + B_s u_{add}(k) \end{cases}. \qquad (2.111)$$

Here, the fault effect must be eliminated in the priority state variables x_p. Thus, from (2.110), this can be achieved by calculating the additive control law u_{add} satisfying

$$(A_{ps} - B_p K_s)x_s(k) + F_{ap}f_a(k) + B_p u_{add}(k) = 0. \qquad (2.112)$$

In this breakdown, if x_s is not available for measurement, it can be computed from the output equation in (2.108), as C_T is a full column rank matrix. Then, the solution u_{add} of (2.112) is obtained using the fault estimation $\hat{f}_a$:

$$u_{add}(k) = -B_p^{-1}[(A_{ps} - B_p K_s)x_s(k) + F_{ap}\hat{f}_a(k)]. \qquad (2.113)$$

The main goal is to eliminate the effect of the fault on the priority outputs. This is realized by choosing the transformation matrix P such that

$$C_T = \begin{bmatrix} C_{T11} & 0 \\ C_{T21} & C_{T22} \end{bmatrix}.$$

Although that the secondary outputs are not compensated for, they must remain stable in the faulty case. Let us examine these secondary variables. Replacing (2.113) in (2.111) yields

$$x_s(k+1) = (A_{ss} - B_s B_p^{-1} A_{ps})x_s(k) - B_s K_2 z(k)$$
$$+ (A_{sp} - B_s K_p)x_p(k) + (F_{as} - B_s B_p^{-1} F_{ap})\hat{f}_a(k). \qquad (2.114)$$

It is easy to see that the secondary variables are stable if and only if the eigenvalues of matrix $(A_{ss} - B_s B_p^{-1} A_{ps})$ belong to the unit circle.

2.7.2 Sensor Fault-tolerant Control Design

As for actuator faults, two main approaches have been proposed to eliminate the sensor fault effect which may occur on the system. One is based on the design of a software sensor where an estimated variable is used rather than the faulty measurement of this variable. The other method is based on adding a new control law to the nominal one.

Sensor Fault Masking

In the presence of sensor faults, the faulty measurements influence the closed-loop behavior and corrupt the state estimation. Sensor FTC can be obtained by computing a new control law using a fault-free estimation of the faulty element to prevent faults from developing into failures and to minimize the effects on the system performance and safety. From the control point of view, sensor FTC does not require any modification of the control law and is also called "sensor masking" as suggested in [131]. The only requirement is that the "estimator" provides an accurate estimation of the system output after a sensor fault occurs.

Compensation for Sensor Faults

The compensation for a sensor fault effect on the closed-loop system can be achieved by adding a new control law to the nominal one:

$$u(k) = -K_1 x(k) - K_2 z(k) + u_{add}(k). \qquad (2.115)$$

It should be emphasized here that, in the presence of a sensor fault, both the output y and the integral error z are affected such that

$$\begin{cases} y(k) = Cx(k) = Cx_0(k) + F_s f_s(k) \\ z(k) = z_0(k) + \tilde{f}(k) \\ \tilde{f}(k) = \tilde{f}(k-1) - T_e F_{s1} f_s(k-1) \end{cases} \qquad (2.116)$$

where x_0 and z_0 are the fault-free values of x and z and $\tilde{f}$ is the integral of $-F_{s1} f_s$. Assuming that matrix $C = I$, these equations lead the control law to be written as follows:

$$u(k) = -K_1 x_0(k) - K_1 F_s f_s(k) - K_2 z_0(k) - K_2 \tilde{f}(k) + u_{add}(k). \qquad (2.117)$$

The sensor fault effect on the control law and on the system can be cancelled by computing the additive control law u_{add} such that

$$u_{add}(k) = K_1 F_s \hat{f}_s(k) + K_2 \tilde{f}(k). \tag{2.118}$$

Remark 2.2. In the case when matrix $C \neq I_n$, the control law can be calculated using the estimated state vector which is affected by the fault as well. The fault compensation will be achieved in a similar way to that given by (2.117) and (2.118).

It has been shown that the new control law added to the nominal one is not the same in the case of an actuator or sensor fault. Thus, the abilities of this FTC method to compensate for faults depend on the results given by the FDI module concerning the decision as to whether a sensor or an actuator fault has occurred.

2.7.3 Fault-tolerant Control Architecture

After having presented the different modules composing a general FTC architecture, the general concept of this approach is summarized in Fig. 2.9 in the linear framework, which is easily extended to the nonlinear case. The FDI module consists of residual generation, residual evaluation, and finally the decision as to which sensor or actuator is faulty. The fault estimation and compensation module starts the computation of the additive control law and is only able to reduce the fault effect on the system once the fault is detected and isolated. Obviously, the fault detection and isolation must be achieved as soon as possible to avoid huge losses in system performance or catastrophic consequences.

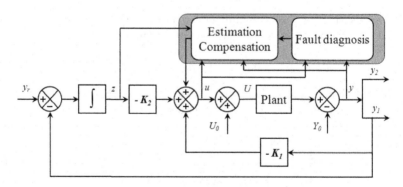

Fig. 2.9. FTC scheme

2.8 General Fault-tolerant Control Scheme

The general FTC method described here addresses actuator and sensor faults, which often affect highly automated systems. These faults correspond to a loss of actuator effectiveness or inaccurate sensor measurements.

The complete loss of a sensor can be overcome by using the compensation method presented previously, provided that the system is still observable. Actually, after the loss of a sensor, the observability property allows the estimation of the lost measurement using the other available measurements. However, the limits of this method are reached when there is a complete loss of an actuator; in this case, the controllability of the system should be checked. Very often, only a hardware duplication is effective to ensure performance reliability.

The possibility and the necessity of designing an FTC system in the presence of a major actuator failure such as a complete loss or a blocking of an actuator should be studied in a different way. For these kinds of failures, the use of multiple-model techniques is appropriate, since the number of failures is not too large. Some recent studies have used these techniques [104, 137, 139].

It is important to note that the strategy to implement and the level of achieved performance in the event of failures differ according to the type of process, the allocated degrees of freedom, and the severity of the failures . In this case, it is necessary to restructure the control objectives with a degraded performance. A complete active FTC scheme can be designed according to the previous classification illustrated in Fig. 1.1. This scheme is composed of the nominal control associated with the FDI module which aims to give information about the nature of the fault and its severity. According to this information, a reconfiguration or a restructuring strategy is activated. It is obvious that the success of the FTC system is strongly related to the reliability of the information issued from the FDI module. In the reconfiguration step, the fault magnitude is estimated. This estimation could be used as redundant information to that issued from the FDI module. The objective of this redundancy is to enhance the reliability of the diagnosis information. The complete FTC scheme discussed here is summarized in Fig. 2.10.

2.9 Conclusion

The FDI and the FTC problems are addressed in this chapter. The complete strategy to design an FTC system is presented. For this purpose, since many real systems are nonlinear, both nonlinear and linear techniques are shown. The linear techniques are used in case the system is linearized around an operating point.

The study presented here is based on the fault detection, the fault isolation, the fault estimation, and the compensation for the fault effect on the system. All these steps are taken into consideration. If this fault allows us to keep

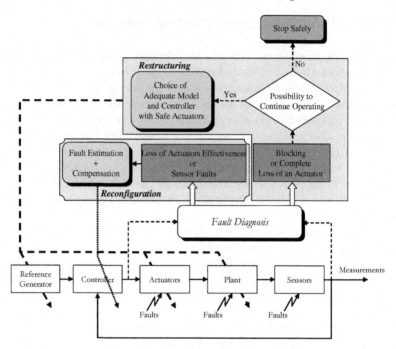

Fig. 2.10. General FTC scheme

using all the sensors and actuators, a method based on adding a new control law to the nominal one is described in order to compensate for the fault effect. For actuator faults, the objective of this new control law is to boost the control inputs in order to keep the performance of the faulty system close to the nominal system performance. Regarding sensor faults, the additive control law aims at preventing the total control inputs from reacting when these faults occur.

In case a major fault occurs on the system, such as the loss of an actuator, the consequences are more critical. This case is analyzed and the system should be restructured in order to use the healthy actuators and to redefine the objectives to reach. Therefore, the system will perform in degraded mode.

The following chapters are dedicated to the application of the linear and nonlinear methods described above to a laboratory-scale winding machine, a three-tank system, and finally in simulation of a full car active suspension system which is considered as a complex system. tured in order to use the healthy actuators and to redefine the objectives to reach. Therefore, the system will perform in degraded mode.

The following chapters are dedicated to the application of the linear and nonlinear methods described above to a laboratory-scale winding machine, a three-tank system, and finally in simulation to a full car active suspension system which is considered as a complex system.

3

Application to a Winding Machine

3.1 Introduction

Web transport systems allow the operations of unwinding and rewinding of various products including plastic films, sheets of paper, sheets, and fabrics. These operations are necessary for the development and the treatment of these products. Web transport systems generally consist of the same machine elements in spite of the diversity of the transported products.

A reduced pilot-plant of an industrial web transport system allows studying of the different operations. The characteristics of this pilot-plant and its control device are presented in the following. The control of these systems is not simple because their dynamic characteristics change throughout unwinding. Moreover, quality standards of rewinding and the requirements of manufacture are often very constraining. The FTC system associated with the fault diagnosis module should improve the control performance without leading to undesirable consequences such as faults or catastrophic breakdowns [124].

A web transport system can be divided into several subsystems (Fig. 3.1). Among them, generally an unroller and a roller are respectively laid out at the beginning and at the end of the process. The other subsystems such as the free rollers dancers, engines tractors, rollers, or the accumulators are used and laid out according to the process of treatment.

The unroller is the starting point of a web transport system. The quality of the reel to be unrolled will influence the behavior of the web during its unwinding. A reel reforming into an oval during storage or badly wound generates disturbances.

A dancer roller downstream from the unroller is used on one hand to impose the unwinding web tension and on the other hand to filter the variations of web tension due to the defects of the reel.

The web is involved and guided by rollers which can be motorized or free. The traction of the web is generally ensured by three rollers: a central roller motorized on a controlled surface and two rollers of smaller sizes, placed so

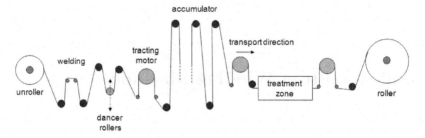

Fig. 3.1. Industrial web transport system

as to obtain a great contact angle web/roller and to support the adherence of the web with respect to the roller tractor.

The accumulator is used to store the web during the normal operations of the system and to restore the web during a change of wind. Consequently it allows a continuous drive.

The step of rolling is the most important among the various operation steps. This step gives the quasi-final aspect to the material and conditions its final quality. The process of rolling in successive layers without appearance of defects is far from easy. Indeed, some folds, some wavelets, or some air pockets can appear and consequently compromise the quality of the product. The causes of these defects are on the one hand, the conditions of rolling (web tension, speed winding, forces support, *etc.*), and on the other hand, the properties of the rolled product (surface topography, homogeneity of elasticity, *etc.*).

The process of web transport must respect several criteria according to the quality of the rolled product. The product should not worsen during its transport because of:

- Elongation
- Crumple
- Folding
- Tear
- Shift

The control of such a system must take into account the characteristics of the product to be transported.

Moreover, web transport systems are multi-variable and coupled systems, the process parameters of which vary during operation. The parameter variation is due to the reels radius variation during unwinding.

The variation of the reels radius significantly modifies the dynamic behavior of the system during the complete process of unwinding. So, the control performances deteriorate. The strategy of industrial control consists in leaving a margin of sufficient stability by reducing the required performances, and more particularly by decreasing the linear velocity of the web.

In this chapter, the objective is to apply the methods described in Chap. 2 to a laboratory-scale winding machine. A complete design of an active FTC system is proposed and analyzed in the presence of minor faults and more critical failures. This is done around an operating point and for a complete unwinding process. The originality of this work is reflected in the description of the effect of various kinds of faults or failures which may affect the system and the classification of FTC techniques according to the severity of these malfunctions.

This study takes into account minor faults and major failures. Minor faults could be biases or drifts on actuators or sensors, a decrease in the actuator effectiveness, or even a complete loss of a sensor under some conditions. Major failures, which involve drastic and discontinuous variations in the plant dynamics, correspond, for instance, to an actuator blocked or out of order. In the presence of such faults, the nominal system performance cannot be reached anymore. Thus, restructuring control objectives with a degraded performance must be set up or the system has to be shut down immediately and safely.

3.2 System Description

3.2.1 Process Description

The winding machine (Fig. 3.2) is composed of three reels driven by DC motors denoted M_1, M_2, and M_3, gears reduction coupled with the reels, and a plastic strip (300 m length, 5 cm broad and 0.2 mm thickness)(Fig. 3.3). The radius of the unwinding reel varies from 210 to 70 mm for the totality of the unwinding operation which lasts approximately 40 min.

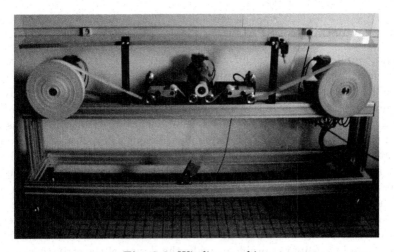

Fig. 3.2. Winding machine

Motor M_1 corresponds to the unwinding reel, M_3 to the rewinding reel, and M_2 to the traction reel. The angular velocity of motor M_2 (Ω_2) and the strip tensions between the reels (T_1, T_3) are measured using a tachometer and tensiometers, respectively. Each motor is driven by a local controller. Torque control is achieved for motors M_1 and M_3, while speed control is realized for motor M_2.

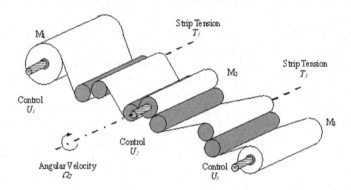

Fig. 3.3. System description

The control inputs of the three motors are U_1, U_2, and U_3. U_1 and U_3 correspond to the current set-points I_1 and I_3 of the local controller as shown in Fig. 3.4. U_2 is the input voltage of motor M_2. In winding processes, the main goal usually consists of controlling tensions T_1 and T_3 and the linear velocity of the strip. Here the linear velocity is not available for measurement, but since the traction reel radius is constant, the linear velocity can be controlled by the angular velocity Ω_2.

3.2.2 Connection to dSPACE®

For the application of a multi-variable control law, a real-time development environment (MATLAB®/Simulink® and Real-Time Workshop® and dSPACE®) is used. The dSPACE® board is the DS1102 based on a Texas Instrument DSP TMS320C31. This board has four analog inputs and four analog outputs. The inputs are the measurements of the strip tensions between the reels (T_1 and T_3), and the angular velocity of motor M_2 (Ω_2). The outputs are the three control inputs U_1, U_2 and U_3 described in Fig. 3.5.

A first study is conducted to illustrate the linear methods developed in Chap. 2. As previously stated, in nominal operation, the objective is to control the strip tensions and velocity. From theoretical point of view, the tension control principle of winding processes seems to be very easy. However, in practice some disturbances such as the vibration, the interaction between the tensions, and the slipping problems make the design of the strip tension control difficult. This winding machine presents one more difficulty: the physical

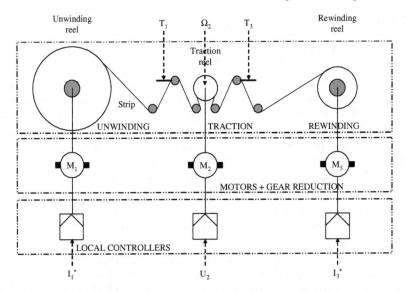

Fig. 3.4. Inputs/outputs of the winding machine

parameters of the system which are time-variant due to the variation of the unwinding and rewinding reel radii. To design a control law for this winding machine, a model describing the nominal behavior is required. This will be described in the next section provided that some assumptions are fulfilled.

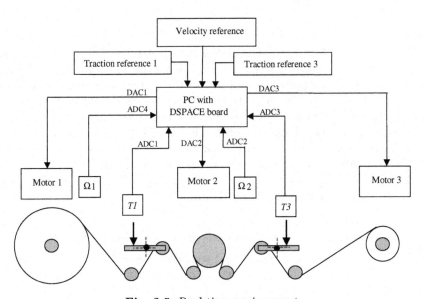

Fig. 3.5. Real-time environment

3.3 Linear Case

3.3.1 System Modeling

Winding processes are generally continuous and nonlinear processes. In practice, the system modeling phase is a tedious and time consuming task and requires a good knowledge of the physical parameters which are often unknown and unmeasurable. Several studies have dealt with the modeling problem of winding systems. Since this is not the main objective of this book, an "average" linear model of the system identified with a nominal working conditions in a small operating zone is considered. For a detailed study of the modeling of this winding machine, the reader can refer to [10].

In the first step, the aim is to consider a linear model around an operating point. Then a multi-linear approach will be considered.

For the linear case, an operating zone is taken around the middle of the strip, $i.e.$, the radius of the unwinding reel is almost equal to the radius of the rewinding. The experiments are then conducted in less than 2 min. This is enough to illustrate the control, FDI, and FTC methods while the model can be still considered as linear.

A "black-box" identification of the winding pilot-plant is used considering a model for each output as a multiple-input single-output (MISO) model. The corresponding analytical model is obtained using an auto-regressive with external inputs (ARX) structure. This model describes the dynamical behavior of the system in terms of input/output variations (u, y) around the operating point (U_0, Y_0):

$$U_0 = \begin{bmatrix} -0.15 & 0.55 & 0.15 \end{bmatrix}^T; \qquad Y_0 = \begin{bmatrix} 0.6 & 0.5 & 0.4 \end{bmatrix}^T. \qquad (3.1)$$

To get this model and to identify the system parameters, pseudo-random binary sequence (PRBS) signals, considered as variations of the control inputs around the operating point, are applied to the system at a sampling period $T_s = 0.1$ s (Fig. 3.6). The magnitude of these signals should be taken such that the system is still operating properly within the operating zone where it is considered as linear. Using the dSPACE® board, the collected signals are given in the interval $\begin{bmatrix} -1 & +1 \end{bmatrix}$ corresponding to $\begin{bmatrix} -10 \ V & +10 \ V \end{bmatrix}$.

The response of the system to these input signals is shown in Fig. 3.7.

After removing the mean value of these signals, and considering the ARX structure, the model for each output is given as

$$\begin{cases} T_1(k+1) = a_{11}T_1(k) + a_{13}T_3(k) + b_{11}u_1(k) + b_{12}u_2(k) + b_{13}u_3(k) \\ \Omega_2(k+1) = a_{21}T_1(k) + a_{22}\Omega_2(k) + a_{23}T_3(k) + b_{21}u_1(k) \\ \qquad\qquad\qquad + b_{22}u_2(k) + b_{23}u_3(k) \\ T_3(k+1) = a_{31}T_1(k) + a_{33}T_3(k) + b_{31}u_1(k) + b_{32}u_2(k) + b_{33}u_3(k) \end{cases} \qquad (3.2)$$

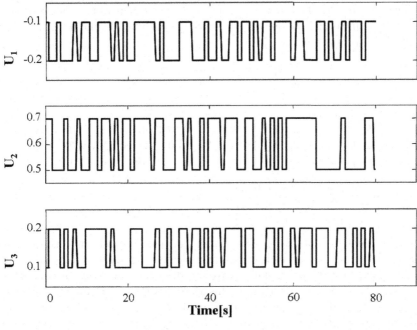

Fig. 3.6. PRBS signals

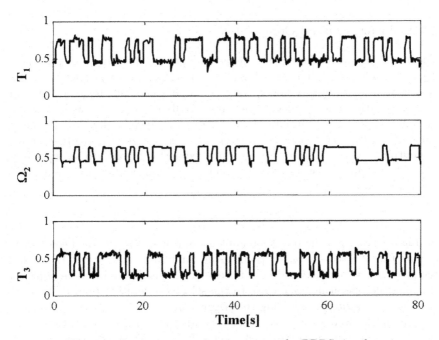

Fig. 3.7. System outputs in response to the PRBS signals

Then, gathering these equations into a discrete-time state-space representation, the linearized model of the winding machine around the operating point (U_0, Y_0) is given by

$$\begin{cases} x(k+1) = Ax(k) + Bu(k) \\ \quad\quad y(k) = Cx(k) \end{cases}, \tag{3.3}$$

with

$$x = \begin{bmatrix} T_1 \\ \Omega_2 \\ T_3 \end{bmatrix}; \quad u = \begin{bmatrix} U_1 \\ U_2 \\ U_3 \end{bmatrix}; \quad A = \begin{bmatrix} 0.4126 & 0 & -0.0196 \\ 0.0333 & 0.5207 & -0.0413 \\ -0.0101 & 0 & 0.2571 \end{bmatrix};$$

and

$$B = \begin{bmatrix} -1.7734 & 0.0696 & 0.0734 \\ 0.0928 & 0.4658 & 0.1051 \\ -0.0424 & -0.093 & 2.0752 \end{bmatrix}.$$

C is the identity matrix I_3. The reader can easily check that the system described by these matrices is completely observable and controllable.

The next step before going on to the design of the control consists of validating this model. To achieve this task, another experiment should be conducted around the same operating point. Then a cross validation is done and the model given by (3.2) and (3.3) is declared to describe the linear behavior of the system around the selected operating point.

3.3.2 Linear Nominal Control Law

The nominal control law is set up according to a tracking control design described in Sect. 2.4. The tracking control problem requires that the number of outputs that have to follow a reference input vector y_r must be less than or equal to the number of control inputs. This is the case for the winding machine: three control inputs U_1, U_2, and U_3 are available, thus the three outputs T_1, Ω_2, and T_3 can be tracked. There is also no need to break down the output vector, we have $y_1 = y$. The feedback controller to set up has to cause the output vector y to track the reference input vector in the sense that in steady-state:

$$y_r(k) - y(k) = 0. \tag{3.4}$$

Figure 3.8 shows the real-time implementation of the designed control law around the selected operating point using Simulink® and dSPACE®. When the program is compiled and runs in real-time, it is driven in open-loop until reaching the operating zone close to the selected operating point. Then, the closed-loop is switched on using the trigger "g_1."

The block "Reference vector" corresponds to the reference vector y_r that the system outputs have to track. Actually, this reference vector contains *variations* of the real references around the operating point Y_0. The block

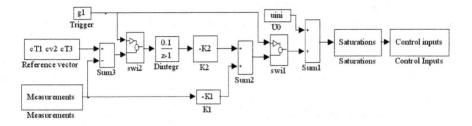

Fig. 3.8. Implementation of the nominal tracking control

"Measurements" corresponds to the data acquisition of the three measurements T_1, Ω_2, and T_3 via the analog inputs of the dSPACE® board (the ADC unit). These three measurements correspond to the output vector Y. These data are filtered using a first order digital filter to reduce the noise level. Then the variations of these measurements around the operating point Y_0 are calculated (Fig. 3.9). These variations correspond to the output vector $y = Y - Y_0$.

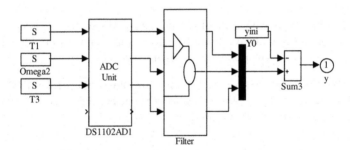

Fig. 3.9. Data acquisition from the sensors

The variations of the control inputs u around the operating points are calculated according to

$$u(k) = - \begin{bmatrix} K_1 & K_2 \end{bmatrix} \begin{bmatrix} x(k) \\ z(k) \end{bmatrix}, \tag{3.5}$$

where z is the integral of the tracking error vector

$$z(k+1) = z(k) + T_s(y_r(k) - y(k)). \tag{3.6}$$

The variations of the control law u (3.5) are added to the operating point values U_0 in order to get the global control law $U = u + U_0$. The saturation of the global control inputs U must be considered before applying it to the actuators via the analog outputs of the dSPACE® (the DAC unit) (Fig. 3.10):

- If $U_1 \geq 1$, then $U_1 = 1$, and if $U_1 \leq -1$, then $U_1 = -1$
- If $U_2 \geq 1$, then $U_2 = 1$, and if $U_2 \leq 0$, then $U_2 = 0$
- If $U_3 \geq 1$, then $U_3 = 1$, and if $U_3 \leq -1$, then $U_3 = -1$

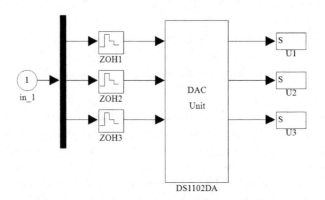

Fig. 3.10. Application of the control inputs

Remark 3.1. The negative values of U_2 mean that the motor is rotating in the opposite sense so U_2 is always equal to zero in our application. The negative values of U_1 and U_3 correspond to a resistant torque.

Remark 3.2. The block "Saturations" and the block "Control inputs" could be combined into one block, but separating them highlights the fact that the saturations should not be forgotten.

Calculation of the Feedback Control Gain Matrix K

After defining the augmented matrices $\bar{A}$ and $\bar{B}$ described by (2.32), $K = \begin{bmatrix} K_1 & K_2 \end{bmatrix}$ is computed using the LQ technique such that the following cost function is minimized:

$$J = \frac{1}{2} \sum_{k=0}^{\infty} \left(X^T(k)QX(k) + u^T(k)Ru(k) \right). \tag{3.7}$$

Weighting matrices Q and R are respectively nonnegative symmetric and positive definite symmetric matrices, $Q = 0.05I_6$ and $R = 0.1I_3$. Using MATLAB®, the feedback gain matrix for the augmented system can be calculated as follows.

```
A_bar = [A zeros(3); -Ts*C eye(3)];
B_bar = [B; zeros(3)];
Q = 0.05*eye(6);
R = 0.1*eye(3);
[K,s,e] = dlqr(A_bar,B_bar,Q,R);
K1 = K(:,1:3);
K2 = K(:,4:6);
```

The application of this control law to the winding machine for a given reference vector close to the operating point leads to the following results illustrated by Figs. 3.11 and 3.12.

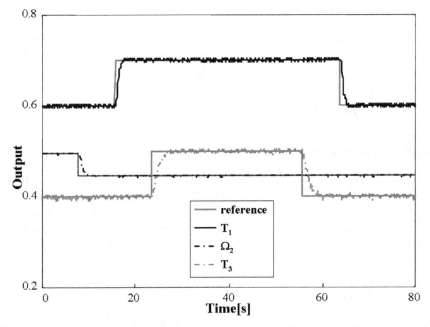

Fig. 3.11. Outputs response of the closed-loop system

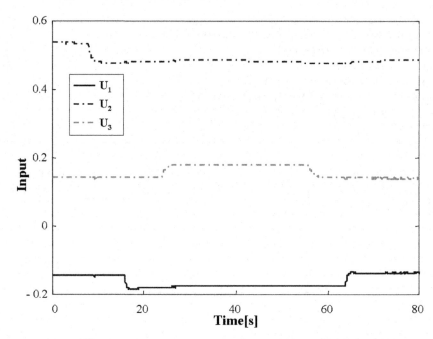

Fig. 3.12. Inputs response of the closed-loop system

3.3.3 Fault-tolerant Control for Actuator Faults

After the design of the nominal control law, the performances of the FTC methods developed in Chap. 2 are tested by simulating actuator and sensor faults according to the description given in Sects. 2.3.1 and 2.3.2. The simulation of the sensor faults consists of modifying the data acquired from the sensors in the Simulink® program, while the simulation of the actuator faults consists of modifying the control inputs to apply to the actuators.

Loss of Effectiveness

As described in Sect. 2.3.1, an abrupt loss of the effectiveness occurring on the ith actuator can be simulated by multiplying the global control input U_i by a constant coefficient $\alpha_i \in \left] 0 \, ; 1 \right[$.

Remark 3.3. To simulate an actuator fault, the global control input U_i should be considered rather than the variation of the control input around the operating point u_i.

Here, an abrupt decrease of 70% in the effectiveness of motor M_3 which is the third actuator has been considered: the control input U_3 is multiplied by $\alpha_3 = 0.3$. Figure 3.13 shows how to simulate this fault in the block "Saturation" of the Simulink® program.

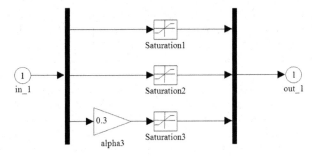

Fig. 3.13. Abrupt loss of the effectiveness of motor M_3

The fault is not considered from the beginning of the experiment. This means that α_3 is equal to 1 at the beginning and when the fault occurs at time t_f, α_3 is switched to 0.3. In the following experiments, the fault occurs at $t_f = 32$ s.

The actuator fault acts on the system as a disturbance. Thus, in the nominal tracking control law considered in this application, the presence of the integrator in the controller compensates for the fault effect in steady-state: the output vector will reach its reference value in steady-state. However, it is obvious that the dynamical behavior of the system will be affected and the system responses will be slower than those in the nominal case.

In order to compensate for this actuator fault effect, the fault magnitude should be estimated and then a new control law u_{add} added to the nominal one. In the following, the application of the fault estimation based on the SVD technique described in Sect. 2.6.3 is presented. Let us recall the system equations in the presence of an actuator fault:

$$
\begin{cases}
\begin{bmatrix} x(k+1) \\ z(k+1) \end{bmatrix} = \begin{bmatrix} A & 0_{3,3} \\ -T_sC & I_3 \end{bmatrix} \begin{bmatrix} x(k) \\ z(k) \end{bmatrix} + \begin{bmatrix} B \\ 0_{3,3} \end{bmatrix} u(k) \\
\qquad\qquad + \begin{bmatrix} 0_{3,3} \\ T_sI_3 \end{bmatrix} y_r(k) + \begin{bmatrix} F_a \\ 0_{3,3} \end{bmatrix} f_a(k) \, . \qquad (3.8) \\
\quad y(k) = \begin{bmatrix} C & 0_{3,3} \end{bmatrix} \begin{bmatrix} x(k) \\ z(k) \end{bmatrix}
\end{cases}
$$

It is possible to consider $F_a = B$; then f_a will be a vector composed of three components. The ith component corresponds to the fault on the ith actuator.

These equations can be rewritten in the following form:

$$\overline{E}_a\overline{X}_a(k+1) = \overline{A}_a\overline{X}_a(k) + \overline{B}_a\overline{U}(k) + \overline{G}_ay_r(k), \qquad (3.9)$$

where

$$\overline{E}_a = \begin{bmatrix} I_3 & 0 & -F_a \\ 0 & I_3 & 0 \\ C & 0 & 0 \end{bmatrix}; \quad \overline{A}_a = \begin{bmatrix} A & 0 & 0 \\ -T_sC & I_3 & 0 \\ 0 & 0 & 0 \end{bmatrix}; \quad \overline{B}_a = \begin{bmatrix} B & 0 \\ 0 & 0 \\ 0 & I_3 \end{bmatrix};$$

$$\overline{G}_a = \begin{bmatrix} 0 \\ T_sI_3 \\ 0 \end{bmatrix}; \quad \overline{X}_a(k) = \begin{bmatrix} x(k) \\ z(k) \\ f_a(k-1) \end{bmatrix}; \quad \overline{U}(k) = \begin{bmatrix} u(k) \\ y(k+1) \end{bmatrix}.$$

The following lines show a section of the MATLAB® program that calculates the necessary matrices to solve the fault estimation problem as described in Sect. 2.6.3.

```
% Actuator  Fault  Estimation
A_bar_a=[A              zeros (3)        zeros (3,3)
         −Ts*C          eye (3)          zeros (3,3)
         zeros (3)      zeros (3)        zeros (3,3)];
B_bar_a=[B              zeros (3)
         zeros (3)      zeros (3)
         zeros (3)      eye (3)];
G_bar_a=[zeros (3)
         Ts*eye (3)
         zeros (3)];

Ea=[eye (3)    zeros (n,p)      −Fa
    zeros (3)  eye (3)          zeros (3)];
Ha=[C      zeros (3)        zeros (3)];

[pea , nea]=size (Ea);
[mea , nea]=size (Ha);

EHa=[Ea;Ha];
[neha , meha]=size (EHa);
% neh = n + p + p,  % meh = n + p + nd

[T1a, S1a , M1a]=svd (EHa);
%   EHa=T1*S1*M1'   % But we need T'*EH*M=S1

Ttransa=inv (T1a); Ma=inv (M1a');
if  S1a (: ,meha)==zeros (neha ,1)
     fprintf ('attention')
end
Sa=S1a (1:meha ,1: meha);

[nTa ,mTa]=size (Ttransa);
```

```
T11a=Ttransa ( 1 : nea , 1 : pea );
T12a=Ttransa ( 1 : nea , pea+1:mTa);
T21a=Ttransa ( nea+1:nTa , 1 : pea );
T22a=Ttransa ( nea+1:nTa , pea+1:mTa);

[ nauga  mauga]= size ( Aauga );  SM_1a=inv ( Sa∗inv (Ma));

Atild_a=SM_1a∗[ T11a  T12a]∗A_bar_a ;
Btild_a=SM_1a∗[ T11aT12a]∗B_bar_a ;
Gtild_a=SM_1a∗[ T11a  T12a]∗G_bar_a ;
```

Once these matrices are calculated, the actuator fault f_a can be estimated by solving the system equations given in (2.98). The real-time actuator fault estimation is added to the Simulink® nominal control program as shown in Fig. 3.14. The additive control law is then calculated according to (2.107).

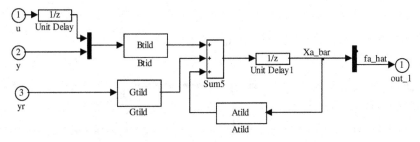

Fig. 3.14. Real-time actuator fault estimation

Application and Results: Actuator Bias Fault

The global actuator FTC is then built as illustrated by the Simulink® program (Fig. 3.15). The complete program is then compiled and executed in real-time for the abrupt actuator fault described above.

The FTC approach enables the compensation for this loss in the dynamical performance of the system outputs corrupted by this fault as illustrated in Fig. 3.16. This fault depends directly on the accommodated input u_3 (Fig. 3.17).

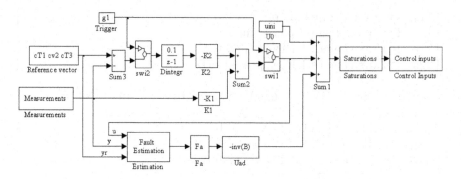

Fig. 3.15. Global actuator FTC scheme

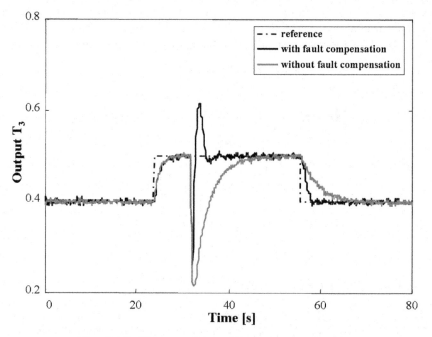

Fig. 3.16. Time evolution of the system output T_3

Figure 3.18 shows the estimation of the faults components associated with each actuator. It can be easily seen that the third component associated with the faulty actuator is only different from zero after the fault occurrence. The estimation of this fault is also compared to its theoretical value after the fault has occurred: $\hat{f}_a = (\alpha_3 - 1)U_3$.

In addition to the visual analysis of the results, the norm of the tracking error $e = y_r - y$ (Table 3.1) can be calculated and analyzed. The norm emphasizes the performances of the FTC method for actuator faults. It can be

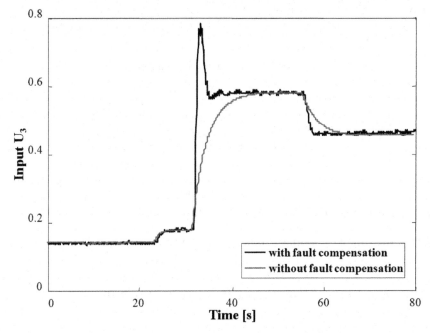

Fig. 3.17. Time evolution of the system input U_3

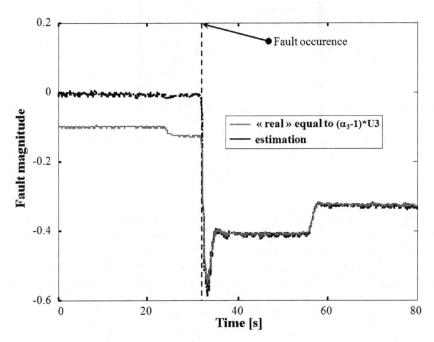

Fig. 3.18. Actuator fault magnitude estimation

easily noted that the norm of the tracking error for the strip tension T_3 after fault compensation is much less than that without fault compensation.

Table 3.1. Norm of the tracking error

	Nominal control	Without compensation	With compensation
$\|e_{T_1}\|$	0.3451	0.3453	0.3506
$\|e_{\Omega_2}\|$	0.1187	0.119	0.1196
$\|e_{T_3}\|$	0.4127	0.7913	0.4692

Application and Results: Actuator Ramp Fault

Then a drift fault is assumed to occur on the third actuator M_3:

$$U_{3f}(k) = U_3(k) - 0.01kT_s. \tag{3.10}$$

This fault can be easily simulated as shown in Fig. 3.19.

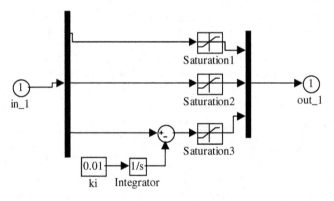

Fig. 3.19. Ramp actuator fault

Although the nominal control law is able to compensate for a constant abrupt fault in steady-state, the presence of a ramp fault leads to a nonzero steady-state error on strip tension T_3. This shows the necessity of the FTC to allow the strip tension to reach its reference value as soon as the fault is detected (Fig. 3.20). However, it is obvious that the compensation for a ramp fault is still possible while maintaining the control input within its physical limits which are -1 and $+1$, corresponding respectively to -10 V and $+10$ V. In this case, the objective of the FTC system is to avoid stopping the system immediately after the fault detection. The fault magnitude estimation is presented in Fig. 3.21.

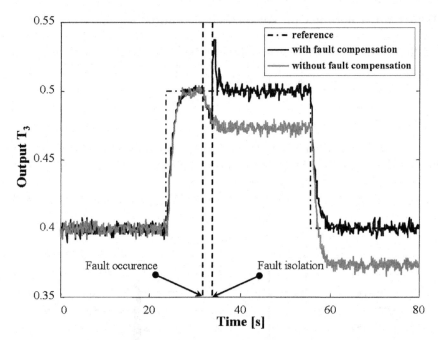

Fig. 3.20. Time evolution of the system output T_3

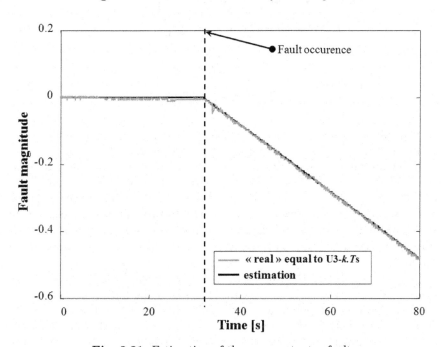

Fig. 3.21. Estimation of the ramp actuator fault

3.3.4 Fault-tolerant Control for Sensor Faults

Let us now examine the influence of sensor faults on the winding machine and the way to compensate for their effect. The sensor faults can appear as a bias, a drift, or a complete loss of the sensor.

Bias Sensor Fault

A first experimentation has been conducted and corresponds to a negative bias on the sensor measuring strip tension T_3 (Fig. 3.22):

$$T_{3f}(k) = T_3(k) - 0.1. \tag{3.11}$$

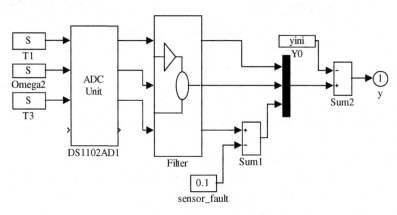

Fig. 3.22. Sensor fault on T_3

The value of the sensor fault is initialized to zero and is switched to -0.1 on-line. In this experiment, the fault occurs at time instant $t_f = 32$ s.

In the presence of this fault, the controller receives the faulty measurement while the true value is still equal to its reference value. The tracking error between the measurement T_{3f} and its reference value is no longer equal to zero. Therefore, the nominal control law tries to bring back the steady-state error to zero. The variation of the control input causes the strip tension T_3 to increase by the same value of the bias on the faulty sensor as it can be seen in Fig. 3.23.

If the bias is larger than 0.1, the real strip tension will be larger which means that the strip may be broken. This shows the importance of taking into consideration the potential sensor fault while designing the control law. Therefore, the control law is computed according to the method presented in Sect. 2.7.2 aiming at preventing the nominal control law from reacting in the presence of this fault. This method is based on the sensor fault estimation

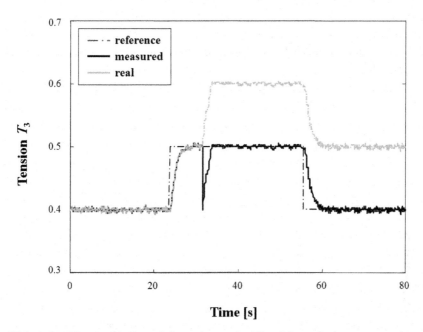

Fig. 3.23. Time evolution of the strip tension T_3 without fault compensation

which is taken into account in the calculation of the control law in order to prevent it from reacting when a sensor fault occurs.

In the presence of sensor faults, the integral error vector z is also affected by the fault:

$$
\begin{aligned}
z(k+1) &= z(k) + T_s(y_r(k) - y(k)) \\
&= z(k) + T_s(y_r(k) - Cx(k) - F_s f_s(k)).
\end{aligned}
\tag{3.12}
$$

The sensor fault magnitude can be estimated in a way similar to the actuator fault, by rearranging the augmented system in the form

$$
\overline{E}_s \overline{X}_s(k+1) = \overline{A}_s \overline{X}_s(k) + \overline{B}_s \overline{U}(k) + \overline{G}_s y_r(k),
\tag{3.13}
$$

where

$$
\overline{E}_s = \begin{bmatrix} I_3 & 0 & 0 \\ 0 & I_3 & 0 \\ C & 0 & F_s \end{bmatrix} ; \quad
\overline{A}_s = \begin{bmatrix} A & 0 & 0 \\ -T_sC & I_3 & -T_sF_s \\ 0 & 0 & 0 \end{bmatrix} ; \quad
\overline{B}_s = \begin{bmatrix} B & 0 \\ 0 & 0 \\ 0 & I_3 \end{bmatrix} ;
$$

$$
\overline{G}_s = \begin{bmatrix} 0 \\ T_sI_3 \\ 0 \end{bmatrix} ; \quad
\overline{X}_s(k) = \begin{bmatrix} x(k) \\ z(k) \\ f_s(k) \end{bmatrix} ; \quad
\overline{U}(k) = \begin{bmatrix} u(k) \\ y(k+1) \end{bmatrix} .
$$

The estimation of the fault magnitude $\hat{f}_s$ is then obtained using the SVD of matrix $\overline{E}_s$.

The estimation of the fault affecting the sensor measuring the strip tension T_3 is shown in Fig. 3.24.

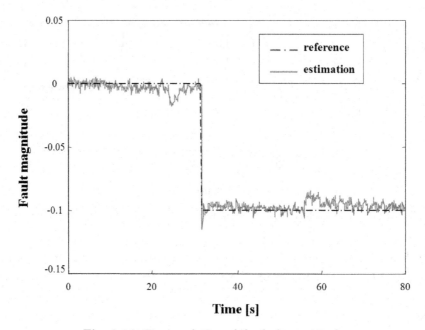

Fig. 3.24. Time evolution of the fault magnitude

The compensation for the sensor fault effect makes use of the sensor fault estimation $\hat{f}_{s3}$. Figure 3.25 shows the nominal control law without taking the fault into account and the new control law able to maintain the real strip tension to its reference value as shown in Fig. 3.26.

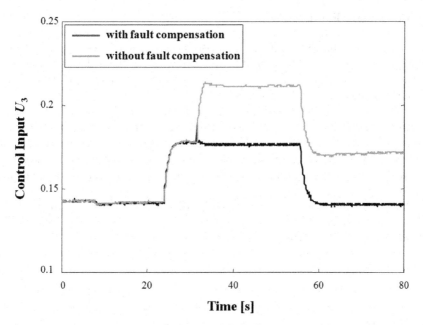

Fig. 3.25. Time evolution of the control input U_3 without and with fault compensation

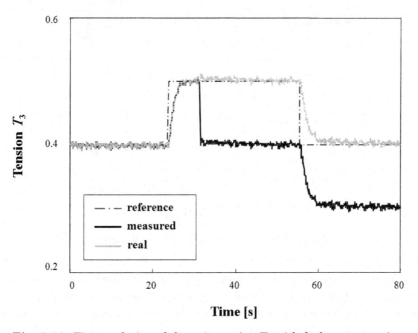

Fig. 3.26. Time evolution of the strip tension T_3 with fault compensation

Ramp Sensor Fault

Incipient faults may also affect the sensors in industrial systems which are mainly due to ageing. These faults cannot always be easily detected because of their slow development. They can be detected when their effect on the system becomes visible. Sometimes this is unacceptable.

The presence of the integrator in the nominal tracking control law leads to a constant nonzero steady-state error between the measurement issued from the sensor and the reference value. At the same time the real measurement deviates slowly from its reference value until the control input reaches the physical limits of the system. In this case the system should be stopped immediately.

An experiment has been conducted by emulating a drift of slope 0.01 on the sensor measuring the strip tension T_3. Figure 3.27 shows the behavior of the strip tension T_3 in the presence of the fault.

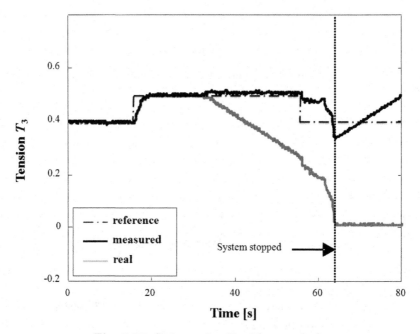

Fig. 3.27. Strip tension T_3 with ramp fault

Using the method presented previously, while taking into account the ramp fault sensor, allows the estimation of the sensor fault magnitude (Fig. 3.28). This estimation is used in order to prevent the control input U_3 from moving face to the fault occurrence as can be seen in Fig. 3.29, which keeps the real value of the strip tension close to its reference value (Fig. 3.30).

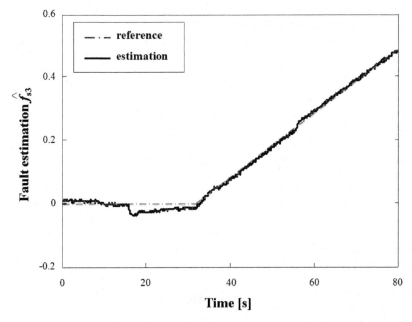

Fig. 3.28. Estimation of the ramp sensor fault

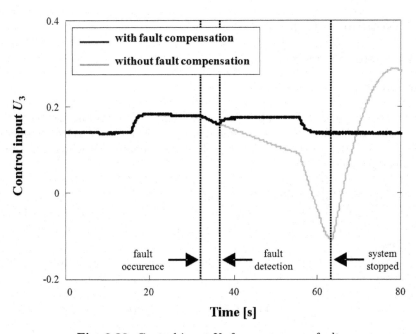

Fig. 3.29. Control input U_3 for ramp sensor fault

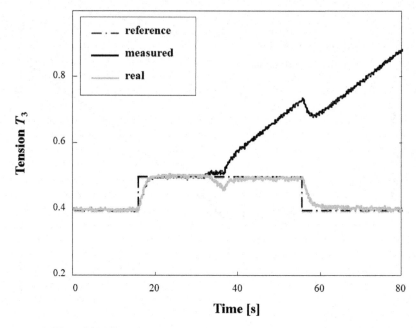

Fig. 3.30. Strip tension T_3 with ramp fault accommodation

3.4 Nonlinear Case

3.4.1 System Modeling

Modeling

During the process of unwinding and rewinding, the radius and inertia of the reels vary. In order to take into account these variations, the system model is defined according to the unwinding reel radius (denoted R) which is a time-varying parameter.

In this case, the dynamic behavior of a winding machine cannot be described using a MIMO linear state-space representation as is done around a particular radius [99, 111]. Indeed, if a complete rolling is performed, the radius variation should be considered and the linear assumption is no longer satisfied.

So the system model can be described by the following nonlinear representation:

$$\xi(k+1) = f(\xi(k)) + g(\xi(k))u(k), \qquad (3.14)$$

where $f(.)$ and $g(.)$ represent the nonlinear functions and $\xi = \begin{bmatrix} T_1 & \Omega_2 & T_3 & R \end{bmatrix}^T$.

Only state variables $\begin{bmatrix} T_1 & \Omega_2 & T_3 \end{bmatrix}$ depend on the control input vector u. The radius is governed by the following equation:

$$R(k) = 0.55(R(k-1) - 1.22\Omega_2(k)). \tag{3.15}$$

Therefore, the system (3.14) can be broken down into two independent subsystems when a specific operating point $y^{ref} = \begin{bmatrix} T_1^{ref} & \Omega_2^{ref} & T_3^{ref} \end{bmatrix}^T$ is considered. State variable Ω_2 is then considered as an input for radius evolution described by (3.15).

Then a discrete linear time-varying (LTV) model, where the time-varying parameter is the unwinding reel radius R (not measured), is obtained such as

$$\begin{cases} x(k+1) = A(R)x(k) + B(R)u(k) \\ \quad y(k) = x(k) \end{cases}, \tag{3.16}$$

where $y = x = \begin{bmatrix} T_1 & \Omega_2 & T_3 \end{bmatrix}^T$, $u = \begin{bmatrix} u_1 & u_2 & u_3 \end{bmatrix}^T$, and $x(k) = x(kT_s)$. T_s is the sampling period.

If an operating point is considered for the unwinding strip, it can be noted that this operating point depends on the nonlinearity introduced by the radius R. The effect of the radius can be considered with the nominal control input (Fig. 3.31a), as well as with the nominal output (Fig. 3.31b).

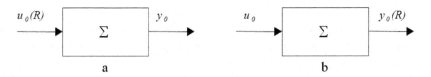

Fig. 3.31. Representation of the operating point

The aim is to control the system outputs, *i.e.*, to control this system around a constant y_0. Thus the representation given in Fig. 3.31a is chosen since the evolution of the outputs does not depend on the radius R.

In order to obtain a model as in (3.16), an identification procedure is established.

Identification

The identification methodology used in this application is composed of four steps:

- The division of the strip into various operating zones (as represented in Fig. 3.32)
- An open-loop identification
- A closed-loop identification
- A polynomial interpolation synthesis

Data organization

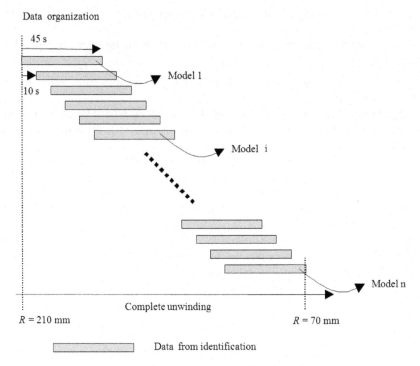

45 s

10 s

Model 1

Model i

Model n

Complete unwinding

$R = 210$ mm

$R = 70$ mm

Data from identification

Fig. 3.32. Data organization for identification

The LTV model (3.16) is obtained after considering 240 operating zones (as represented in Fig. 3.32), where each operating zone can be modeled with an LTI model. Experiments of 45 s for each operating zone are used in order to consider an LTI representation; $i.e.$, the radius is considered to be constant (variation between 1 and 4 mm) and is manually measured (no sensor is available), with a sampling period $T_s = 0.1$ s.

The second step, dedicated to the open-loop identification of each operating zone, following the same methodology as presented in the linear case, allows to obtain 240 analytical models using a classical ARX structure [90,99]. In order to identify each LTI model parameter, a PRBS is added as an external input to the nominal control input $u_{nom} = \begin{bmatrix} -0.05 & 0.4 & 0.15 \end{bmatrix}^T$ (the system inputs and outputs are given in the interval $\begin{bmatrix} -1 & +1 \end{bmatrix}$ corresponding to $\begin{bmatrix} -10 \ V & +10 \ V \end{bmatrix}$).

The measured outputs vary with the radius R of the strip roll over the whole operating range as shown in Fig. 3.33.

The final goal is to design a control law for angular velocity Ω_2 and strip tensions T_1 and T_3. Therefore, the system must be characterized for constant measured outputs as proposed in Fig. 3.31a. Thus, an elementary and classical LQ with integrator control law [45] is synthesized and implemented for each

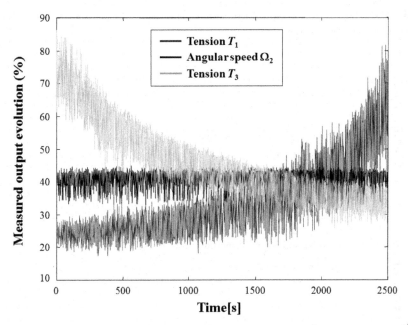

Fig. 3.33. Measured outputs with constant inputs $u_{nom} = \begin{bmatrix} -0.05 & 0.4 & 0.15 \end{bmatrix}^T$

LTI model identified above (*i.e.*, for each operating zone). Since there is an integrator in the control law, the measured outputs are constant and equal to the reference inputs which are chosen to be equal to $y_{nom} = \begin{bmatrix} 0.2 & 0.4 & 0.5 \end{bmatrix}^T$.

The third step consists of a closed-loop identification of an LTI model for each operating zone by means of a direct solution [30]. It can be noted that the results are biased since the plant input signal u is correlated with the output noise disturbance and because no noise model is specified. But this solution is kept in the first approach. It can also be noted that since the controller is completely known for each LTI model, an indirect solution could also be applied.

The last step is the acquisition of one LTV model. The 240 LTI models are considered and a polynomial interpolation, dependent on the radius R, of each component of matrices A and B has been synthesized to approximate the dynamic behavior of the LTV system. Then, each coefficient in these matrices is expressed in the following form ($\forall\, i,j = 1, \cdots, 3$):

$$\begin{aligned}
\{a_{ij}, b_{ij}\}R(k) = p_{ij}^0 + p_{ij}^1(R(k)) + p_{ij}^2(R(k))^2 + p_{ij}^3(R(k))^3 \\
+ p_{ij}^4(R(k))^4 + p_{ij}^5(R(k))^5 + p_{ij}^6(R(k))^6,
\end{aligned} \tag{3.17}$$

where $p_{ij}^\sigma (\sigma = 0, \cdots, 6)$ are constant values of polynomial form.

An LTV model described by (3.16) was established and is considered in the following. Each coefficient of $A(R)$ and $B(R)$ matrices is defined by an optimal sixth order polynomial function of the radius R. Figures 3.34 and 3.35 validate the previous results.

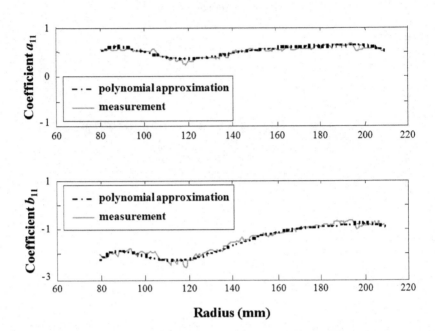

Fig. 3.34. Evolutions of model coefficients a_{11} and b_{11} vs radius R

Figure 3.34 compares the measured and approximated evolutions of two coefficients of the model according to the radius R (coefficient a_{11} of matrix A and coefficient b_{11} of matrix B). Note that the polynomial approximation is very close to the measurements.

Figure 3.35 represents the evolution of the eigenvalues of the system according to the radius (one real and two complex conjugate eigenvalues). The crosses represent the eigenvalues associated with each of the 240 identified models and the curve represents the eigenvalues of the variable system according to R. The eigenvalues are very close in both approximations.

3.4.2 Controller Gain Synthesis

The winding process model presented previously is linear with a variable parameter and is available around the operating point (U_0, Y_0). Moreover, the number of measured outputs is equal to the number of control inputs. This model is completely suitable to design a control law using nonlinear techniques by linearization presented in Sect. 2.4.2. In this section, the input-output linearizing control, developed within the framework of the continuous systems is considered in discrete-time.

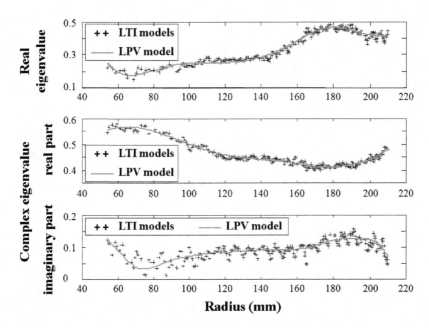

Fig. 3.35. Representation of the eigenvalues of the system as function of R

The definitions and the theorem of Sect. 2.4.2 can be applied to the LTV model of the winding machine (3.16) with $f(x) = A(R)x(k)$, $g(x) = B(R)$, and $h(x) = Cx(k)$ [102].

According to the methodology described in Sect. 2.4.2, the first step corresponds to the calculation of the relative degrees. All are equal to 1. The linearizing state-feedback is thus written as

$$u(k) = -B^{-1}(R)A(R)x(k) + B^{-1}(R)v(k). \tag{3.18}$$

Note that this feedback is nonlinear since it depends on R. $B^{-1}(R)$ can be verified off-line in order to guarantee the stability of the system. The linearized system is then decoupled and three SISO subsystems can be considered for the description of the input-output behavior, each one being equivalent to an exact delay such as

$$y_i(k+1) = v_i(k), \qquad \forall i \in [1,\ldots,3]. \tag{3.19}$$

Such a closed-loop system requires a second state-feedback, built using linear control theory. A proportional output feedback v_i is set up for each decoupled subsystem to perform this task. Consequently, each discrete input-output z-transfer function is given by

$$\frac{y_i(z)}{y_{ref,i}(z)} = \frac{(1-K_i)z}{z-K_i}, \qquad \forall i \in [1,\ldots,3], \tag{3.20}$$

where K_i allows adjustment of the stability dynamic and $y_{ref,i}$ is the reference input.

The closed-loop system can be represented by Fig. 3.36. In this study, all the states are available through the measurements; thus no observer is required to estimate the states.

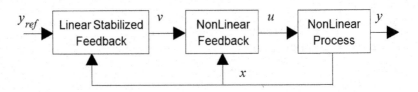

Fig. 3.36. Block diagram of the nonlinear control

The control law described above is valid if and only if the model of the considered process is exact. In the presence of modeling errors, which is generally the case in practice, some steady-state errors will appear on the controlled outputs. The modeling errors can be considered as a disturbance input d. Thus, (3.16) can be written in the following form:

$$\begin{cases} x(k+1) = A(R)x(k) + B(R)(u(k) + d(k)) \\ \quad\quad y(k) = x(k) \end{cases} . \tag{3.21}$$

In order to reject modeling errors, a disturbance rejection control is used as proposed in [103]. In the presence of this disturbance input d, and with the linearizing state-feedback (3.18), the subsystem $\frac{y_i}{v_i}$ will not be strictly equivalent to r_i. A component $v_{add,i}$, which will be added to v, must be determined in order to verify (3.19). Theoretical output $\widehat{y}(k)$ obtained in the presence of an *exact* decoupling, is considered and estimated. For the winding system, this estimation is obtained via a classical Luenberger observer of a system equivalent to an exact delay, such as

$$\widehat{y}_i(k) = v_i(k-1) + L(y_i(k-1) - \widehat{y}_i(k-1)), \tag{3.22}$$

where L is the observer gain.

The observation error vector obeys the relation

$$\varepsilon_i(k+1) = -L\varepsilon_i(k) - v_{add,i}(k), \tag{3.23}$$

where $\varepsilon_i(k) = y_i(k) - \widehat{y}_i(k)$, $\forall i \in [1,\ldots,3]$.

So, each component of v_{add} is defined by

$$v_{add,i}(k) = -(\varepsilon_i(k+1) + L\varepsilon_i(k)). \tag{3.24}$$

Equation (3.24) associated with the disturbance rejection depends on the future values at time instant $(k+1)$. Consequently, a delay is introduced to establish this recurrence.

The complete control law can be illustrated by Fig. 3.37.

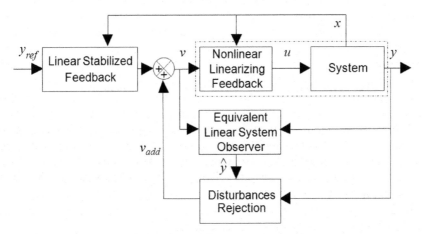

Fig. 3.37. Complete control scheme

Figure 3.38 shows the Simulink® program of the designed nonlinear control law. In order to implement it and test it using dSPACE® software, the "Winding Machine" block must be replaced by input and output ports connected to the input-output card.

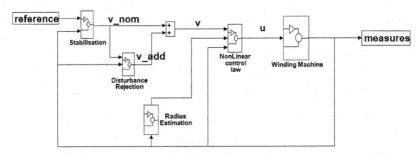

Fig. 3.38. Implementation of the nonlinear control law

Figure 3.39 shows the details of the disturbance rejection block associated with (3.22)–(3.24).

In the following tests, the various signals are only represented on part of the complete unwinding strip for legibility.

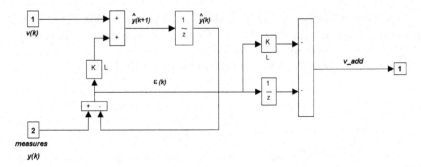

Fig. 3.39. Details of the disturbance rejection implementation

First, the decoupling control law (3.18) is validated in simulation (as presented in Figs. 3.40–3.42 using the discrete-time LTV model estimated in Sect. 3.4.1. Steps of magnitude 30% are added successively to the reference inputs. It can be noticed that the outputs follow the reference inputs and that they are decoupled from each other.

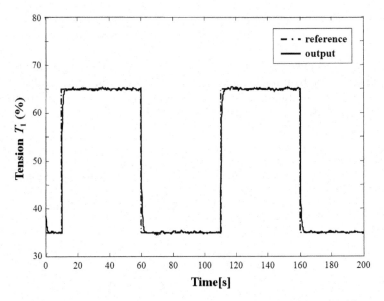

Fig. 3.40. Simulated output T_1 based on the linearizing control law

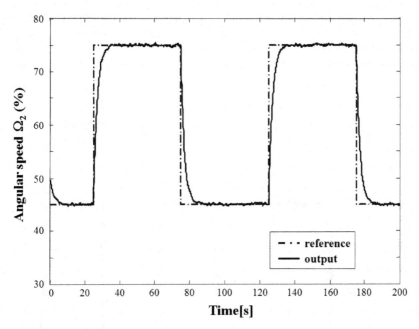

Fig. 3.41. Simulated output Ω_2 based on the linearizing control law

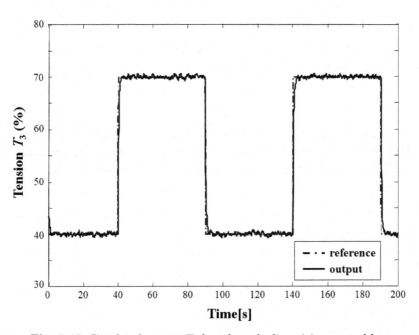

Fig. 3.42. Simulated output T_3 based on the linearizing control law

Figures 3.43–3.45 check the integral action added to the control of the real system. Indeed, it can be verified that the steady-state errors are canceled. Moreover, the outputs are decoupled. Nevertheless, a significant noise affects the measurements.

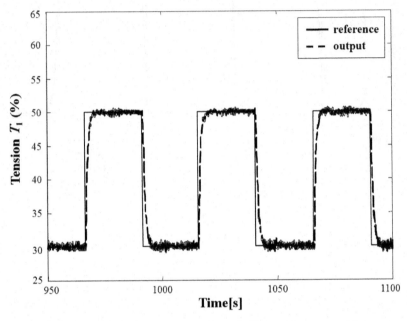

Fig. 3.43. Real output T_1 based on the linearizing control law

The experiment illustrated by Figs. 3.46–3.48 highlights the exact decoupling approach associated with the reduction of the noise level (compared to Figs. 3.43–3.45, respectively). This result is obtained using a Kalman filter [55]: the control law is computed with these estimations instead of sensor measurements.

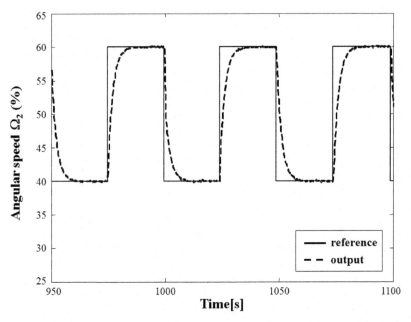

Fig. 3.44. Real output Ω_2 based on the linearizing control law

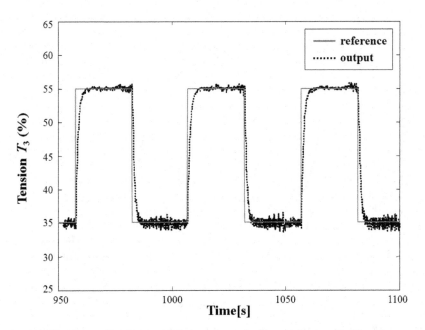

Fig. 3.45. Real output T_3 based on the linearizing control law

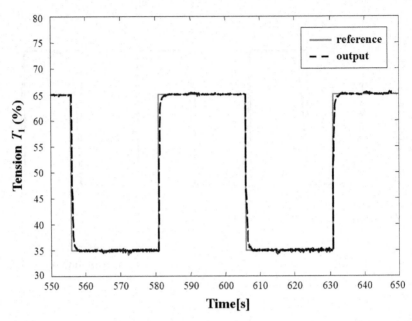

Fig. 3.46. Control law performances with a Kalman filter: measured output T_1

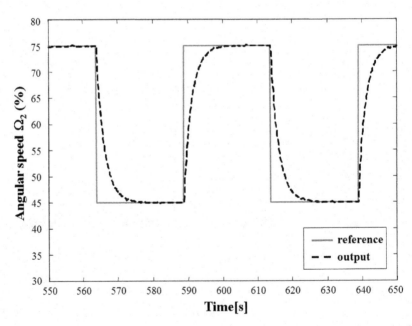

Fig. 3.47. Control law performances with a Kalman filter: measured output Ω_2

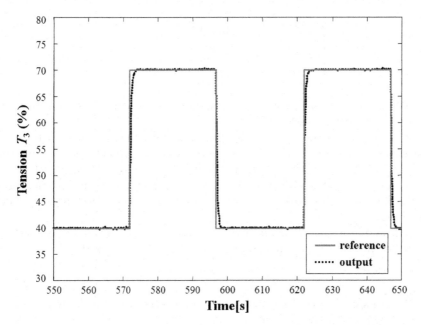

Fig. 3.48. Control law performances with a Kalman filter: measured output T_3

3.4.3 Fault-tolerant Control for Actuator Faults

In order to detect actuator faults with the detection filter (2.84) presented in Chap. 2, the winding process model (3.16) becomes

$$\begin{cases} x(k+1) = A(R)x(k) + B(R)u(k) + Ff_k(k) \\ y(k) = x(k) \end{cases}, \qquad (3.25)$$

where $F = B(R)$.

For the winding machine, the fault detectability indexes are equal to one for all faults and regardless of the value of radius R.

The linear detection filter (2.84) is extended and applied to the LTV model (3.25).

In order to facilitate the parametrization of the FDI module, a bank of detection filters is considered. Each filter is dedicated to detect and to isolate a unique actuator fault. Equation (3.25) is rewritten as

$$\begin{cases} x(k+1) = A(R)x(k) + B(R)u(k) + F_i f_i(k) \\ y(k) = x(k) \end{cases}, \qquad (3.26)$$

where F_i is the ith column of matrix $B(R)$.

Let us define the matrices for each detection filter in the case of the considered application. From (2.85) and (2.86), the gains of the detection filters (2.84) are defined as

$$\begin{cases} \omega_i = \omega_i(R) = A(R)F_i \\ \Xi_i = \Xi_i(R) = (CF_i)^+ \\ \Psi_i = \Psi_i(R) = \beta \left(I_3 - (CF_i)(CF_i)^+ \right) \end{cases} \qquad (3.27)$$

The gain K is synthesized by a common eigenstructure assignment such that $(A(R) - K)$ is stable.

The fault detection is directly established with the fault magnitude estimation vectors $\hat{f}_i(k)$ from the $\eta(k)$ component of the projected residual vector $p(k)$ as presented in (2.89). Each residual is associated with each fault magnitude estimator. Then residuals have directional properties in response to a particular fault. This kind of residuals is an attractive way for enhancing fault isolation.

The actuator fault compensation is implemented according to the method described in Sect. 2.7.1. Thus, for the Winding Machine, (2.107) becomes

$$u_{add}(k) = - \left(B^{-1}(R) \right) F_i \hat{f}_i(k). \qquad (3.28)$$

As in nominal case, $B^{-1}(R)$ should be checked off-line in order to guarantee the stability of the closed-loop.

Figure 3.49 illustrates the Simulink® diagram allowing the validation of the presented method on the Winding Machine model. Each block corresponds to the differential equations developed previously such as "actuator isolation filter," "fault estimation," and "actuator fault compensation."

First, FDI and fault estimation modules effectiveness are considered in both simulation and real experiment. Then, FTC results will be illustrated by considering a significant fault on actuator M_2.

To show the efficiency of the FDI module, actuator faults are carried out starting from levels of magnitude 2% added to the inputs of the system which is controlled with the nonlinear and decoupling control law. These faults are biases which appear and disappear after a short period. The first one concerns the input of motor 1 (U_1), the second one affects the input of motor 2 (U_2) and the last fault occurs on the input of motor 3 (U_3).

First, FDI module is validated in simulation. Noise was added on the system outputs.

In the case of an exact LTV model, the fault detection, isolation and estimation module indicates which actuator is faulty and estimates accurately the fault magnitude. The estimated faults are decoupled from each other and they have the same magnitude as the faults applied to the system. In fault-free cases, residuals have an average value close to zero.

The results are presented in Figs. 3.50–3.52 around $R = 210 \ mm$ and are identical for the whole unwinding strip regardless of the value of the radius.

The results for bias type faults in a complete unwinding process on the real system are presented in Fig. 3.53 (the unwinding reel is initialized with a radius of 210 mm and then with 70 mm). The three actuator faults have

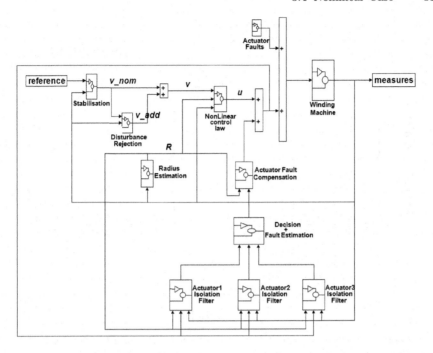

Fig. 3.49. FTC scheme for the winding machine

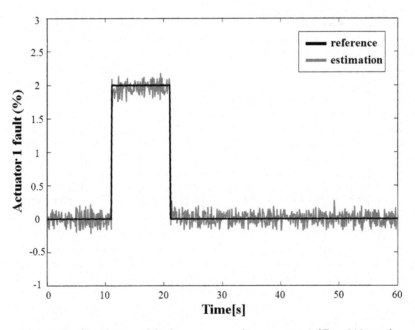

Fig. 3.50. Simulation of fault estimation for actuator 1 ($R = 210\ mm$)

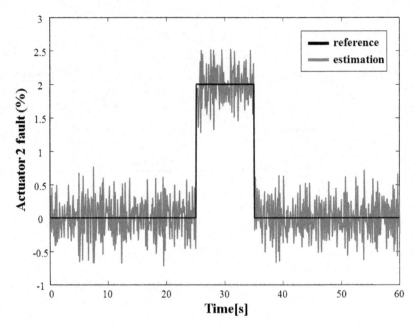

Fig. 3.51. Simulation of fault estimation for actuator 2 ($R = 210\ mm$)

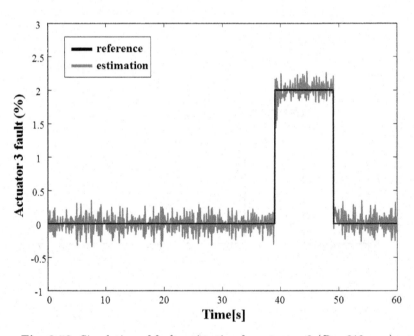

Fig. 3.52. Simulation of fault estimation for actuator 3 ($R = 210\ mm$)

the same magnitude as for the simulations presented above and are applied
to the process with a square form during all the unwinding process.

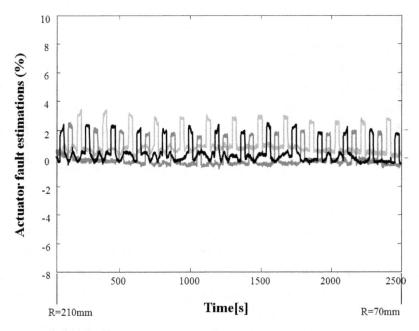

Fig. 3.53. Estimated actuator faults for a complete unwinding

For legibility, Figs. 3.54–3.56 represent the results displayed on Fig. 3.53
around the radius value equal to 157 *mm*.

As for the simulation test presented above, it can be noted that actuator
faults are well detected and isolated. When there is no fault, the fault estimation has a constant average value close to zero. From a practical point of
view, this knowledge should be taken into account in the residual evaluation
part.

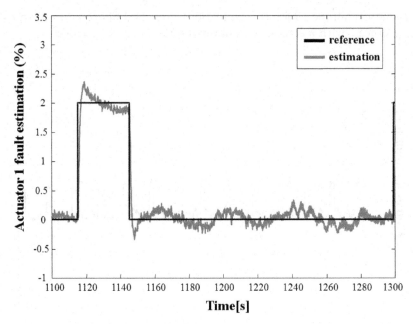

Fig. 3.54. Estimated actuator 1 fault ($R = 157\ mm$)

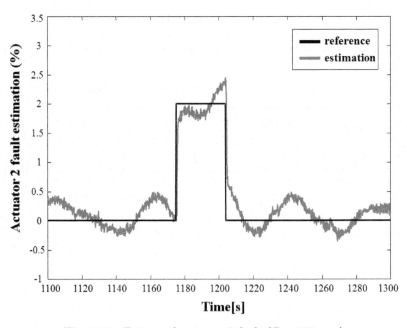

Fig. 3.55. Estimated actuator 2 fault ($R = 157\ mm$)

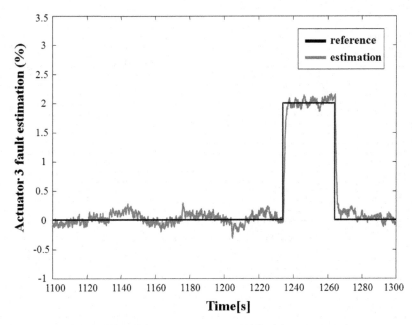

Fig. 3.56. Estimated actuator 3 fault $(R = 157\ mm)$

In order to illustrate the FTC results, another test is carried out in a second experiment. The effectiveness of the second actuator M_2 acting on the strip velocity is reduced by 50% and appears abruptly for a radius equal to 150 mm at time instant 1175 s. During complete unwinding, steps of magnitude ± 0.1 are added successively to the reference inputs.

The system outputs are displayed in Fig. 3.57 without accommodating the control law and with the accommodation.

The actuator fault acts on the system as a disturbance. Since there is an integral action (via the disturbance rejection module) associated with the nonlinear control law, the fault effect in steady-state can be compensated for: the output vector will reach its reference value in steady-state. However, it is obvious that the dynamical behavior of the system will be affected and the system responses will be slower than those in the nominal case.

In order to compensate for this actuator fault effect, a new control law u_{add} is added to the nominal one as specified in (3.28) and in Fig. 3.49.

Figure 3.58 displays the component $\eta(k)$ of the projected residual vector $p(k)$ (2.89) which corresponds to the estimation of the fault magnitude of actuator 2.

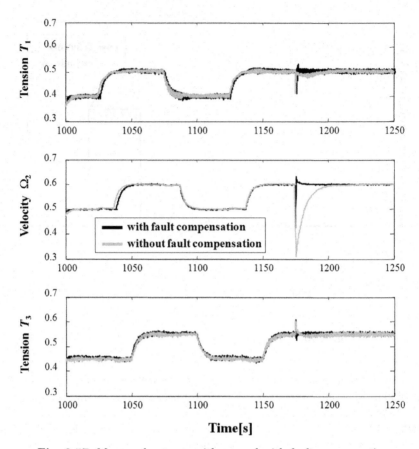

Fig. 3.57. Measured outputs without and with fault compensation

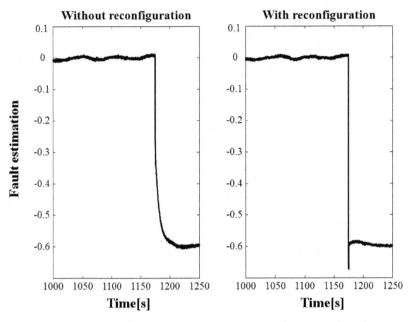

Fig. 3.58. Estimated fault magnitude without and with reconfiguration

The display of measurements during a short period of time presented in Fig. 3.59 allows the comparison of the system responses to the faults without and with fault compensation. These figures clearly demonstrate the FTC method's ability to compensate for such actuator faults.

The results obtained clearly show the ability of the residuals to detect and isolate faults. Once the fault is isolated with a time delay, the corresponding fault estimation and compensation module is switched on to reduce the fault effect on the system through the additive control input u_{add}.

It is shown that, without FTC, the velocity reaches its corresponding reference input about 20 s after the fault occurrence, whereas it takes only about 2 s using the FTC method.

These results can be confirmed by examining the control inputs (Figs. 3.60 and 3.61) applied to the system: without the FTC method, it increases slowly due to the integral error trying to compensate for the fault effect, whereas the FTC method makes this control input increase quickly and enables the rapid fault compensation, even if a coupling effect can be noted on control input U_1.

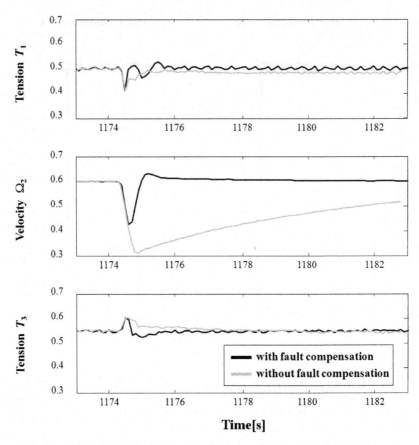

Fig. 3.59. Measured outputs without and with reconfiguration

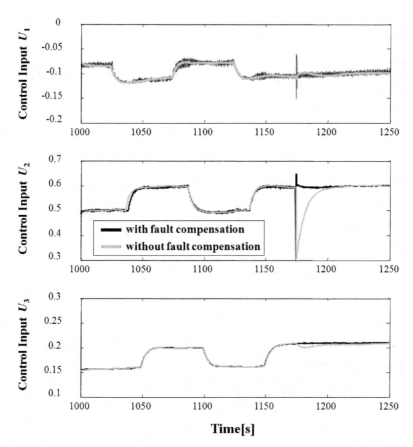

Fig. 3.60. System control inputs without and with reconfiguration

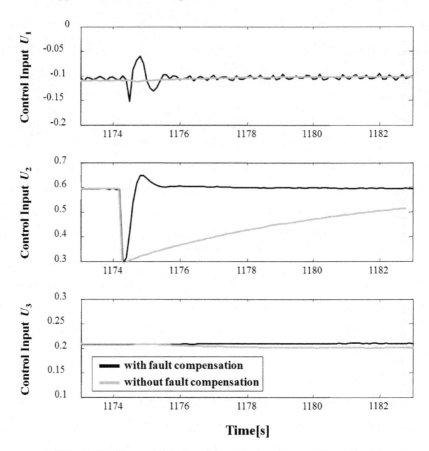

Fig. 3.61. Control inputs without and with reconfiguration

3.4.4 Fault-tolerant Control for Sensor Faults

Based on the capability to represent a sensor fault as a pseudo-actuator fault (see Chap. 2), the previous filter allows us to compute a magnitude estimation of sensor faults. Before using the filter, a preliminary work consists of rewriting LTV faulty model using Park *et al.* developments [100] as detailed in Sect. 2.5.1. The fault detection is directly established with the magnitude fault estimation vectors. Each residual is associated with each fault magnitude estimator. Then residuals have directional properties in response to a particular fault. This kind of residuals is an attractive way for enhancing fault isolation.

A first experiment is considered with software sensor faults to each sensor except Ω_2 which is considered to estimate the radius $R(k)$ as presented in the previous paragraph. A bias with a magnitude of 10% appears and disappears on the real process during all the unwinding strip. For legibility, results for a radius value close to 180 mm are presented in Figs. 3.62 and 3.63 where the real and estimated fault magnitudes highlight the capabilities of the FDI module. Actually, based on LTV model, the fault detection, isolation, and estimation module indicates which sensor is faulty and estimates accurately the fault magnitude. In a fault-free case, residuals have an average value close to zero.

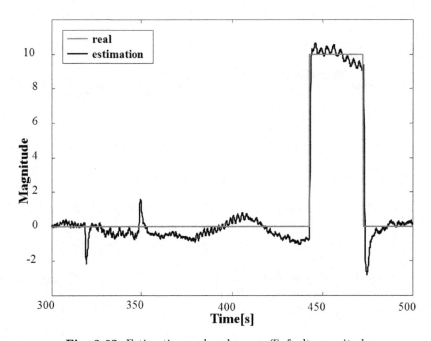

Fig. 3.62. Estimation and real sensor T_1 fault magnitude

The next experiment considers a sensor fault only on the measured output T_1. A bias with a magnitude of 0.05, which corresponds to step changes of the reference inputs equal to 50%, is applied throughout all the unwinding strip. For legibility, magnified images for a radius value around 150 mm are presented.

Based on the LTV model, the fault detection, isolation, and estimation module indicates which sensor is faulty and estimates accurately the fault magnitude as presented in Fig. 3.64.

Without sensor fault masking, the control law tries to cancel the steady-state error created by the corrupted output. Consequently, the real output is different from the reference input (Fig. 3.65). As illustrated in Fig. 3.66, the synthesized nonlinear control law provides decoupled outputs despite the occurrence of a sensor T_1 fault.

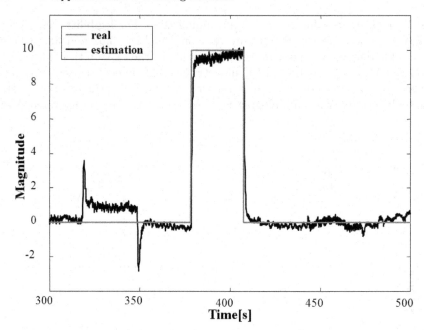

Fig. 3.63. Estimation and real sensor T_3 fault magnitude

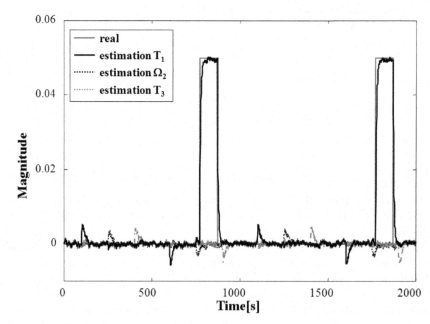

Fig. 3.64. Sensor fault magnitude estimations around $R = 150\ mm$ with a bias on T_1 without sensor fault masking method

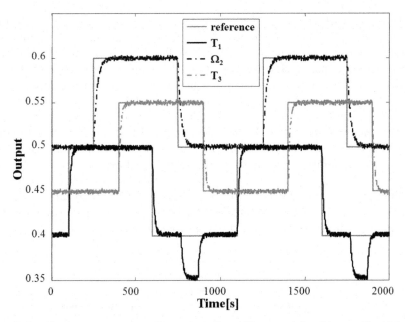

Fig. 3.65. Real system output T_1 around $R = 150$ mm with a bias on T_1 without sensor fault masking method

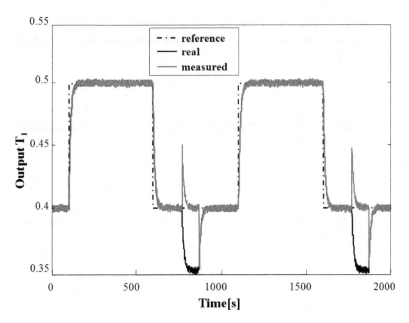

Fig. 3.66. Real and measured system outputs around $R = 150$ mm with a bias on T_1 without sensor fault masking method

However, with the developed sensor fault compensation method, the real output follows its reference input as illustrated in Fig. 3.67. The fault estimation is used to track the reference input. The dynamic behavior of the real and measured outputs shows clearly that the compensated output behavior is closer to the nominal one than the faulty output without compensation as illustrated in Fig. 3.68. As presented in Fig. 3.69, the real output is only affected by the fault according to the fault isolation time delay. In this case the time delay is very small due to the considered fault magnitude compared to the noise level. The sensor fault compensation avoids faults developing into failures and minimizes the effects on the system performance and safety. The proposed FDI strategy represents an efficient tool in the operator's decision winding process. This sensor fault monitoring should be associated with the sensor FTC system to provide alarms about the new dynamic behavior of the system concerning, for instance, the measured outputs.

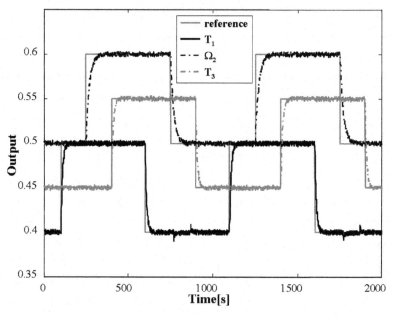

Fig. 3.67. Real system outputs around $R = 150\ mm$ with sensor fault masking method in the presence of a bias on T_1

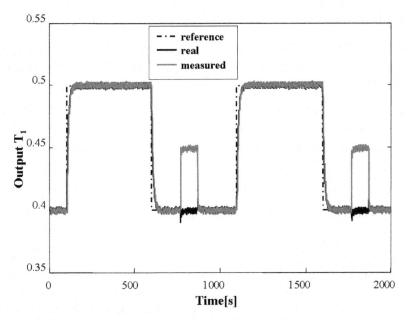

Fig. 3.68. Real and measured system output T_1 around $R = 150\ mm$ with sensor fault masking method in the presence of a bias on T_1

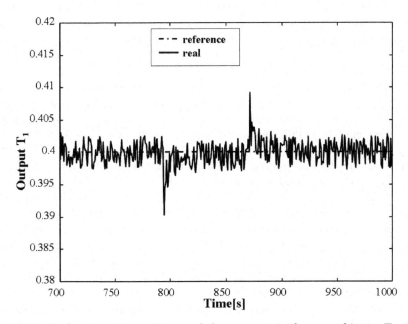

Fig. 3.69. Real system output T_1 around the occurrence of a sensor bias on T_1 with sensor fault masking method

3.5 Major Fault

In the previous two sections, biases or drifts affecting the actuators or the sensors have been considered. In the sequel, the FTC of the winding machine will be studied when a major fault occurs. A complete loss of an actuator or a sensor are considered as a major fault. It is easily understood that, without an FTC law, the winding machine cannot continue operating: system shutdown is necessary.

Next, the situation of a major actuator fault will be studied in the linear case according to Sect. 3.3. Then, in the nonlinear case, the complete loss of one sensor will be considered according to Sect. 3.4.

3.5.1 Actuator Failure in Linear Case

For bias or drift faults, it was shown that it is still possible to track all the outputs of the system, because the latter is supposed to remain controllable in the presence of an actuator fault. This was achieved using another control law, added to the nominal one, in order to compensate for the faults. This additive control law is based on the estimation of the fault magnitude. The reconfiguration strategy employed here allows us to maintain the performance of the faulty system in closed-loop close to the performance of the nominal system; the tracking control objectives remain achieved. However, the limits of this method are reached in case of a stuck or completely lost actuator. Now, the possibility to continue operating with degraded performance is analyzed in the presence of a critical failure such as the complete loss of an actuator.

In the winding machine, the strip has to be rolled up in a correct way, $i.e.$, strip tensions T_1 and T_3 have to be maintained to a given reference level. In this application, this is achieved by maintaining a negative (resistant) torque on motor M_1 and a positive torque on motor M_3. Due to the strong coupling in this system, the angular velocity Ω_2 of the traction reel M_2 influences the strip tensions.

Here motor M_1 is supposed to be out of order at time instant $t_f = 49.5\ s$; $i.e.$, motor M_1 runs as if its control input $U_1 = 0$. This failure leads to a large decrease in strip tension T_1 which can no longer be controlled. With the nominal control law, the coupling in the system causes angular velocity Ω_2 to increase trying to compensate for T_1; Ω_2 increases about 20%. Not only is it impossible to compensate for T_1, but this also has the opposite effect on strip tension T_3, which decreases almost 20% from its reference value. That makes the strip roll up incorrectly. Figures 3.70–3.73 illustrate the failure effect on the system inputs and outputs.

For this specific failure where the motor M_1 is lost, the system operates badly in terms of the product quality but not in terms of the system stability or security. It is also easy to understand that if this failure occurs on one of the other motors M_2 or M_3, the system will no longer be able to run and should be stopped safely. That is to say that it is not always possible to accommodate

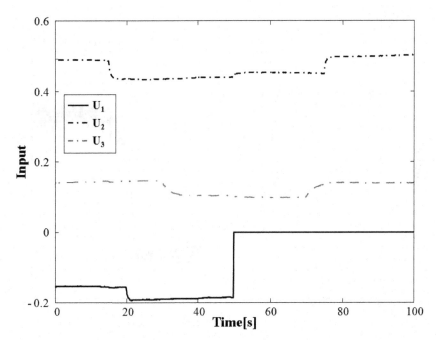

Fig. 3.70. Control inputs when M_1 is lost

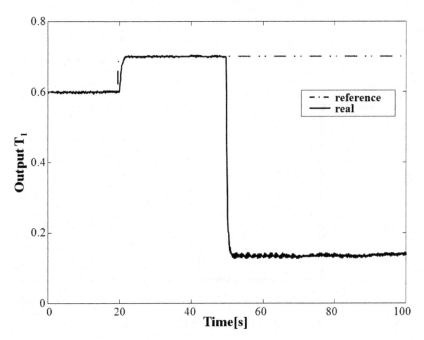

Fig. 3.71. Strip tension T_1 when M_1 is lost

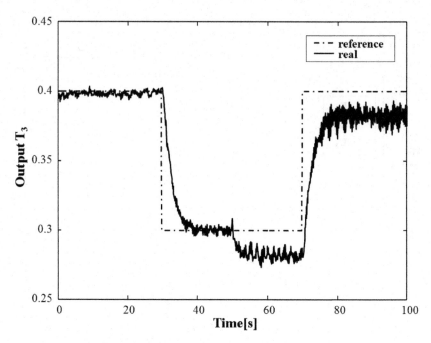

Fig. 3.72. Strip tension T_3 when M_1 is lost

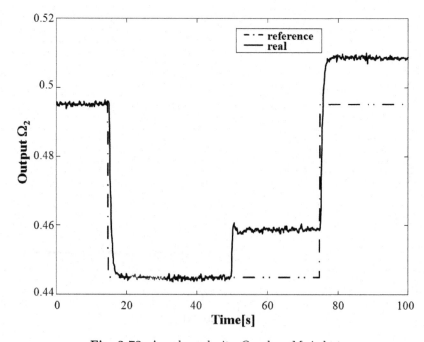

Fig. 3.73. Angular velocity Ω_2 when M_1 is lost

for all failures occurring on the system. It always depends on the available hardware or analytical redundancy.

We have seen that, if motor M_1 is lost, it is still possible to continue operating but the performance is reduced. Now the question is: is it possible to do something to preserve the main performance and keep a good quality of the product despite the loss of this motor.

This is a severe failure because it leads to a large loss in the closed-loop system performance. As one of the system control inputs is out of order, it becomes impossible to track the three system outputs because the tracking control requires the number of control inputs to be greater than or equal to the number of outputs to be tracked. Hence, according to the system operation requirements, these outputs have to be divided into priority outputs to keep equal to their reference inputs to the detriment of other secondary outputs. From the knowledge of the system operating conditions, although one control input is no more available (namely M_1), the system is still controllable. Analyzing the behavior of the system, it is easy to select the output to take off. Strip tension T_3 should be controlled to roll up the strip properly. Angular velocity Ω_2 is the image of the linear velocity of the strip and needs to be kept controlled, while there is no loss in the quality or the productivity if strip tension T_1 is no longer controlled. Thus, the priority outputs will be T_3 and Ω_2, while T_1 will be considered as a disturbance. The system control inputs are U_2 and U_3.

There is no hardware redundancy available in the winding machine. One can easily understand that it is very difficult, if not impossible, to design a hardware redundancy for such a system. Moreover, if the nonlinear model of the system was available, it would have been possible to calculate a new equilibrium state of the system the closest possible to the nominal system, but the nonlinear model of the winding machine is very difficult to get. This point will be illustrated in the next chapter for the three-tank system.

In the case that the nonlinear model of the system is not available, one way to cope with these critical failures is to get a model of the faulty system. For these kinds of critical failures, the number of faulty models is limited. From the knowledge that one has of the system, it is easy to define which failures could be compensated for and which require the shutdown of the system immediately and safely.

In the winding machine there are three actuators driven by control inputs U_1, U_2, and U_3. Motor M_2 has to impose the velocity of the strip. If this motor is out of order, it is obvious that the strip cannot be rolled up and the system must be shut down immediately. It is also the case if motor M_3 is out of order, because strip tension T_3 can no longer be maintained to its reference value. Thus, the only critical failure to deal with in this system is the loss of motor M_1.

Faulty System Model and Results

Since two control inputs are only available after the failure has occurred, it is impossible to continue tracking the three system outputs. Therefore, a tracking control law using the same principle as described previously, based on the faulty system model, has been achieved to track two system outputs Ω_2 and T_3 which are considered as priority outputs. It has been noted previously that strip tension T_1 is considered as a disturbance. With this system restructuring, the identification of the faulty system has been achieved off-line. The system is running with $U_1 = 0$, and a set of excitation signals, U_2 and U_3, has been applied to the winding machine. Using these experiments, the following model is obtained:

$$\begin{cases} x_f(k+1) = A_f x_f(k) + B_f u_f(k) + B_d T_1(k) \\ \qquad y_f(k) = C_f x_f(k) \end{cases}, \qquad (3.29)$$

where

$$x_f = y_f = \begin{bmatrix} \Omega_2 \\ T_3 \end{bmatrix}; \qquad u_f = \begin{bmatrix} U_2 \\ U_3 \end{bmatrix},$$

and

$$A_f = \begin{bmatrix} 0.4995 & -0.0262 \\ 0.0723 & 0.4314 \end{bmatrix}; B_f = \begin{bmatrix} 0.53 & 0.0236 \\ -0.0782 & 1.2865 \end{bmatrix}; B_d = \begin{bmatrix} 0.2149 \\ -0.0490 \end{bmatrix}.$$

Once the failure is detected and isolated, the FTC module switches from the nominal control law to the new one. This control law guarantees the fact that the strip continues to be rolled up properly and avoids having to stop the machine due to a bad quality of the final product.

Figures 3.74–3.77 show the results obtained when switching from the original model to the new one after the failure has been detected and isolated. The FDI module is not achieved here, but a delay of 10 sampling periods is considered before switching to the new control law. This delay corresponds to the detection and isolation task. It can be seen that strip tension T_1 is still far from its reference value because it is not tracked, while strip tension T_3 and the angular velocity Ω_2 follow their respective reference inputs after the switching process and the strip is rolled up in a correct way.

It was shown that for more critical failures, such as a complete loss of an actuator, where the system becomes uncontrollable, the system has to be restructured and new objectives have to be defined. These objectives correspond, for instance, to the modification of the reference inputs and the system outputs to control. Here, the number of outputs to track is reduced. In case it is possible to design such an FTC method for this kind of major failure, the corresponding model is achieved off-line. According to the process described by Fig. 2.10, the supervision module at the upper level decides to switch to the accurate control law according to the isolated failure. In the system used here, there is no problem of security, but the system is still able to continue its operation avoiding loss in productivity and quality of the product. It should

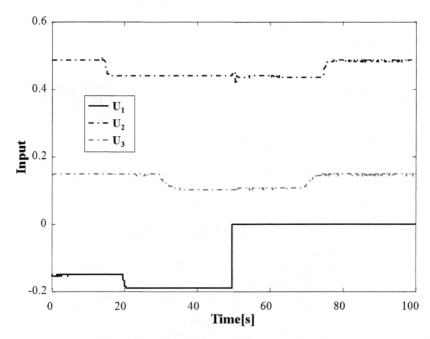

Fig. 3.74. Control inputs after restructuring

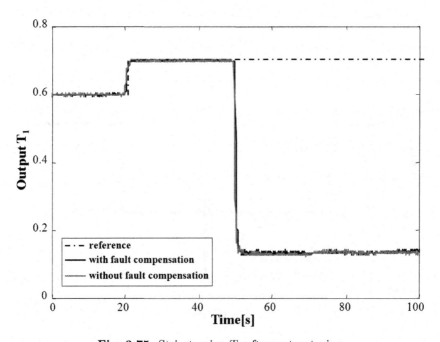

Fig. 3.75. Strip tension T_1 after restructuring

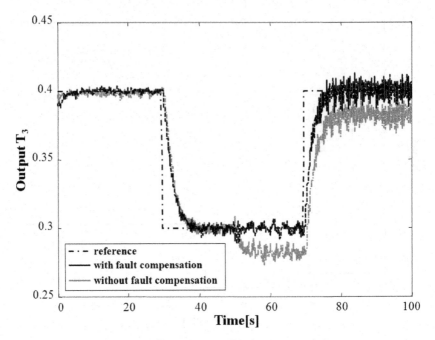

Fig. 3.76. Strip tension T_3 after restructuring

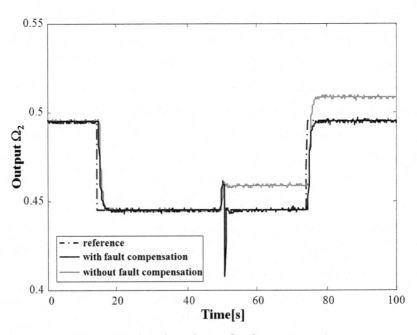

Fig. 3.77. Angular velocity Ω_2 after restructuring

be noticed that, whatever the importance of the FTC system designed, it is never obvious to compensate for all kinds of failures. In this case, the only possible solution is to shut down the system safely.

3.5.2 Complete Loss of a Sensor in Nonlinear Case

The last experiment concerns a complete loss of measurement Ω_2 when considering the nonlinear approach. The radius $R(k)$ is not measured directly but estimated through the recursive equation $R(k) = 0.55(R(k-1) - 1.22\Omega_2(k))$. Due to the fault occurrence on Ω_2, radius $R(k)$ cannot be estimated correctly. In this experiment, the complete loss of sensor Ω_2 appears from time instant $t = 50$ s to $t = 100$ s, then from sample $t = 250$ s to $t = 300$ s and finally from sample $t = 450$ s to $t = 500$ s. As illustrated in Fig. 3.78, the estimation of radius R is corrupted. When sensor $\Omega_2(k)$ is out of order (*i.e.*, $\Omega_2 = 0$ V), the linear evolution of the radius time function is affected by the fault.

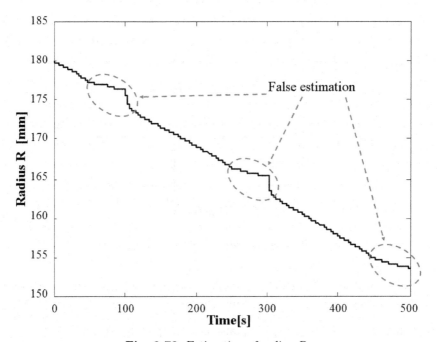

Fig. 3.78. Estimation of radius R

In the presence of such a failure, the faulty measurement influences the closed-loop behavior associated with the desired reference input but also the LTV model used for the decoupled control law.

A test was carried out for a radius $R(k)$ varying between 180 mm and 150 mm with step changes (20% of their corresponding operating values) added to the nominal reference inputs. Figures 3.79 and 3.80 represent, respectively, the measured and real controlled output responses to reference changes (step responses are considered for a range of 150 s). The complete loss of the sensor can easily be seen in the second graph in Fig. 3.79. The control law tries to cancel the steady-state error created by the corrupted output. Consequently, the real output differs from the reference input as illustrated in the second graph in Fig. 3.80. Regarding the MIMO controller, the effect of this loss of information has some drastic consequence on the other tracking outputs as shown in the other graphs of Fig. 3.80.

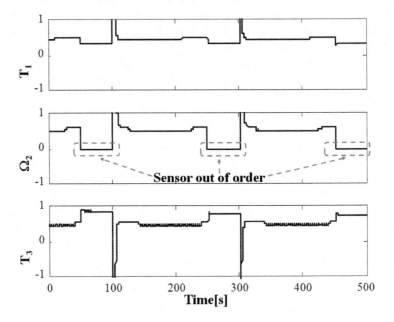

Fig. 3.79. Measured outputs evolution (T_1, Ω_2, T_3)

Based on the LTV model, the fault detection, isolation, and estimation module indicates which sensor is faulty and provides an accurate fault magnitude estimation as illustrated in Fig. 3.81. It can be verified that the first and third components of the fault magnitude estimator are close to zero. These components are used to isolate the faulty sensor. Due to the detectability index of the filter, it should be highlighted that the fault magnitude estimator associated with the first and third components is affected by an abrupt variation different from zero during one sample when the fault appears and also disappears. Indeed, the detectability index of the filter synthesized for a sensor fault is equal to one which provides a time delay of one sample.

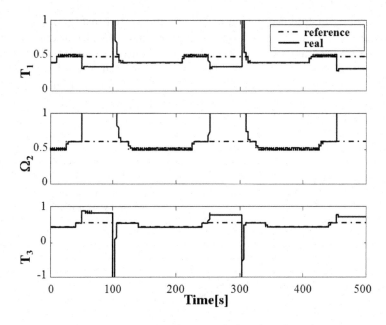

Fig. 3.80. Real outputs evolution (T_1, Ω_2, T_3)

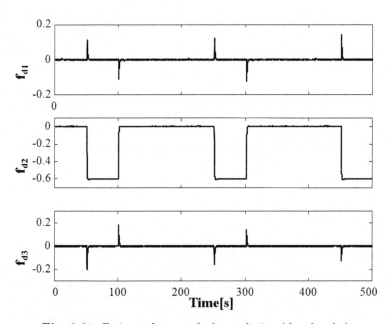

Fig. 3.81. Estimated sensor faults evolution $(f_{d_1}, f_{d_2}, f_{d_3})$

Based on the FDI module, a suitable estimation of radius R is provided through the fault-free estimation of Ω_2. As illustrated in Fig. 3.82, the estimation of $R(k)$ is not affected by the loss of sensor Ω_2. Consequently, with the sensor fault compensation method, the real outputs follow their reference inputs as presented in Fig. 3.83. The fault estimation is used to track the reference input. According to the time delay issued from the FDI module, the results show that the compensated outputs behavior is closer to the nominal outputs rather than the faulty outputs.

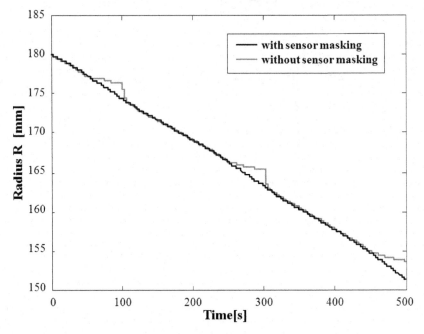

Fig. 3.82. Estimation of radius R with sensor fault compensation

3.6 Conclusion

A winding machine representing a subsystem in many industrial systems is presented in this chapter. It is used to illustrate the FTC strategy presented in the previous chapter. Since the system is nonlinear and the mathematical model is difficult to establish, it is first considered to be linear around an operating point. The model is obtained experimentally using identification methods. Then, in order to run the system throughout a wider operating zone, the model is taken as an LTV system driven by the radius of the unwinding reel.

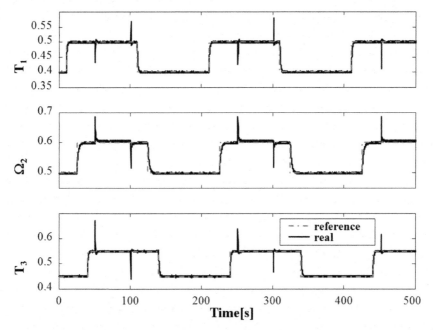

Fig. 3.83. Real outputs (T_1, Ω_2, T_3) with sensor fault compensation

The effect of sensors and actuators faults are considered. It was shown that the methods using the additive control law are able to compensate for the fault effect provided that the system is still observable in the case of a sensor fault and controllable in the case of an actuator fault. The fault estimation can also be used to detect and isolate the fault. Indeed, while the fault estimation is close to zero in the fault-free case, it deviates from zero and it is equal to the fault magnitude when a fault occurs. The system can be written such that only the ith component of the fault vector f_a or f_s deviates from zero when the ith actuator or sensor is faulty.

For major failures such as the loss of an actuator, it is obvious that the nominal tracking control is no longer suitable. This is because the number of control inputs becomes less than the number of outputs to track. In this case, in addition to the understanding of the physical behavior of the system, an off-line study helps one to select priority outputs to keep tracking to the detriment to other secondary outputs. This restructuring strategy allows one to avoid the immediate shutdown of the system and to keep operating in degraded mode.

4

Application to a Three-tank System

4.1 Introduction

In this chapter, a hydraulic system that can be used for water treatment or storing liquids in many industrial plants is considered. During these processes, chemical reactions are supposed to occur around pre-defined operating points. Therefore, the liquid levels control in a plant is crucial in order to provide desired specifications. Using a prototype of a hydraulic system, researchers have successfully tested various methods of linear or nonlinear decoupling control and model-based fault diagnosis.

The three-tank system considered in this study is a popular laboratory-scale system designed by [4]. It is used in order to investigate linear, nonlinear multivariable feedback control as well as FDI and FTC system design. Koenig *et al.* [82] have synthesized a decoupled linear observer to detect and to isolate actuator and component faults (pipe, tank, etc.) around an operating point without fault magnitude estimation. Based on the nonlinear model, [112] have designed an observer using the bilinear model representation of the three-tank system to detect a leakage from a pipe. A diagnostic system based also on a bilinear model has been considered in [7], where time varying innovation generators combined with generalized likelihood ratio tests are designed to detect and isolate faults. The robustness of a sliding mode observer (SMO) to detect faults in the presence of noise on the measurements was tested in real-time [105]. Among model-based approaches, a differential geometric method has been successfully applied in [76]. Rather than considering a complex nonlinear model, [1] have estimated the state vector based around various operating points through a bank of decoupled observers to generate residuals for fault detection. More recently, [106] have proposed to develop a bank of decoupled observers to detect and isolate actuator/sensor faults around multiple operating points applied to the three-tank system. FDI methods based on fuzzy or neural models have also been illustrated on the three-tank system to detect and isolate faults [83, 91, 93]. In [115] the authors deal with the FDI problem of plants with unknown description.

In the FTC framework, the three-tank system has been considered as a benchmark. In the presence of sensor faults, [143] estimates the fault magnitude based on an adaptive filter with an on-line parameter estimation method developed by [142]. The sensor fault estimation is used for sensor fault masking. In [25] an effective low-order tuner for FTC of a class of unknown nonlinear stochastic sampled-data systems is proposed. The strategy is based on the modified state-space self-tuning control via the observer/Kalman filter identification method. Weighted fuzzy predictive control is used for FTC of an experimental three-tank system [95]. Furthermore, a European project "COSY" (control of complex systems) [62] has considered the three-tank system as a benchmark for all partners under the assumption: two tanks are active and the last one is used as a redundant process.

While various FDI and FTC approaches in the literature have been applied separately to the three-tank system, this chapter aims at presenting a complete approach in order to present and illustrate the application of the methods developed in Chap. 2 to this system. This is investigated in the linear case around an operating point as first presented by [121] but also on the whole operating zone using nonlinear techniques. A complete simulation platform of the three-tank system, in closed-loop, with or without actuator and sensor faults, is provided for the use of the reader via download from www.springer.com/978-1-84882-652-6 as described in the Appendix.

4.2 System Description

The considered hydraulic system is presented in Fig. 4.1.

Fig. 4.1. Three-tank system

The hydraulic system consists of three identical cylindrical tanks with equal cross-sectional area S (Fig. 4.2). These three tanks are connected by two cylindrical pipes of the same cross-sectional area, denoted S_p, and have the same outflow coefficient, denoted μ_{13} and μ_{32}. The nominal outflow located at tank 2 has the same cross-sectional area as the coupling pipe between the cylinders but a different outflow coefficient, denoted μ_{20}. Two pumps driven by DC motors supply the first and last tanks. Pumps flow rates (q_1 and q_2) are defined by flow per rotation. A digital/analog converter is used to control each pump. The maximum flow rate for pump i is denoted q_{imax}. A piezo-resistive differential pressure sensor carries out the necessary level measurement. Three transducers deliver voltage signal levels. The variable ℓ_j denotes the level in tank j and ℓ_{jmax}, the associated maximum liquid level.

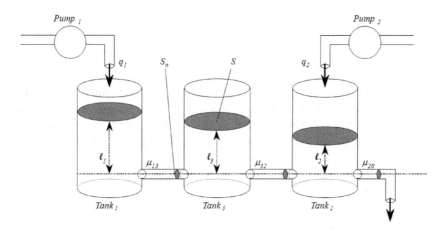

Fig. 4.2. Synoptic of the three-tank system

The system can be described by the following mass balance equations:

$$
\begin{cases}
S\dfrac{d\ell_1(t)}{dt} = q_1(t) - q_{13}(t) \\[2mm]
S\dfrac{d\ell_2(t)}{dt} = q_2(t) + q_{32}(t) - q_{20}(t) \, , \\[2mm]
S\dfrac{d\ell_3(t)}{dt} = q_{13}(t) - q_{32}(t)
\end{cases}
\tag{4.1}
$$

where q_{mn} represents the flow rate from tank m to n ($m, n = 1, 2, 3 \; \forall m \neq n$) which, based on the Torricelli law, is equal to

$$
q_{mn}(t) = \mu_{mn} S_p sign(l_m(t) - l_n(t)) \sqrt{2g \mid l_m(t) - l_n(t) \mid}, \tag{4.2}
$$

and q_{20} represents the outflow rate described as follows:

$$
q_{20}(t) = \mu_{20} S_p \sqrt{2g l_2(t)}. \tag{4.3}
$$

The numerical values of the plant parameters are listed in Table 4.1.

Table 4.1. Parameters value of the three-tank system

Variable	Symbol	Value
Tank cross sectional area	S	$0.0154\ m^2$
Inter tank cross sectional area	S_p	$5 \times 10^{-5}\ m^2$
Outflow coefficient	$\mu_{13} = \mu_{32}$	0.5
Outflow coefficient	μ_{20}	0.675
Maximum flow rate	$q_{imax}(i \in [1\ 2])$	$1.2 \times 10^{-4}\ m^3 s^{-1}$
Maximum level	$l_{jmax}(j \in [1\ 2\ 3])$	$0.62\ m$

4.3 Linear Case

4.3.1 Linear Representation

Under the assumption $\ell_1 > \ell_3 > \ell_2$, a linear model can be established around an equilibrium point (U_0, Y_0). The system is linearized around this operating point using Taylor expansion. The linearized system is described by a discrete LTI representation with a sampling period $T_s = 1\ s$:

$$\begin{cases} x(k+1) = Ax(k) + Bu(k) \\ \quad y(k) = Cx(k) \end{cases}, \tag{4.4}$$

where y and u represent variations around an operating point defined by the pair (U_0, Y_0).

The purpose of this study is to control the system around the operating point (U_0, Y_0), which is fixed to

$$\begin{cases} Y_0 = \begin{bmatrix} 0.40 & 0.20 & 0.30 \end{bmatrix}^T (m) \\ U_0 = \begin{bmatrix} 0.35 \times 10^{-4} & 0.375 \times 10^{-4} \end{bmatrix}^T (m^3/s) \end{cases}. \tag{4.5}$$

In order to generate matrices A and B, the following program is written in MATLAB® code:

```
% Parameters value of three-tank system
    mu13=0.5; mu20=0.675; mu32=0.5;
    S=0.0154; Sn=5e-5; W=sqrt(2*9.81);

% Output operating Points (m)
    Y10=0.400; Y20=0.200; Y30=0.300;

% Input operating Points (m3/s)
    U10=0.350e-004; U20=0.375e-004;

% Matrix A
    A11=-(mu13*Sn*W)/(2*S*sqrt(Y10-Y30)); A12=0;
    A13=-A11;
```

```
A21=0;A23=(mu32*Sn*W)/(2*S*sqrt(Y30-Y20));
A22=-A23-((mu20*Sn*W)/(2*S*sqrt(Y20)));
A31=-A11;A32=A23;A33=-A32-A31;
A=[A11 A12 A13; A21 A22 A23; A31 A32 A33];

% Matrix B
B11=1/S;B12=0;
B21=0;B22=1/S;
B31=0;B32=0;
B=[B11 B12;B21 B22;B31 B32];

% Continuous to discret state space transformation
[Ad, Bd] = c2d(A,B,1.0);
```

Then, matrices A, B and C are equivalent to

$$A = \begin{bmatrix} 0.9888 & 0.0001 & 0.0112 \\ 0.0001 & 0.9781 & 0.0111 \\ 0.0112 & 0.0111 & 0.9776 \end{bmatrix} ; \quad B = \begin{bmatrix} 64.5687 & 0.0014 \\ 0.0014 & 64.2202 \\ 0.3650 & 0.3637 \end{bmatrix} ; \quad C = I_{3 \times 3}.$$

Remark 4.1. As presented in Chap. 3, which discussed the winding machine application, (4.4) may be obtained using an identification method.

4.3.2 Linear Nominal Control Law

A tracking control problem is considered in this study where the desired outputs $y_1 = [\ell_1 \ \ell_2]^T$ are required to track references y_r with

$$y(k) = \begin{bmatrix} y_1(k) \\ y_2(k) \end{bmatrix} = \begin{bmatrix} C_1 \\ C_2 \end{bmatrix} x(k), \tag{4.6}$$

where $y_2 = \ell_3$.

To achieve the nominal tracking control, the solution proposed by [29] and developed in Sect. 2.4.1 has been considered for the three-tank system. Since the feedback control can only guarantee the stability and the dynamic behavior of the closed-loop system, a complementary controller is required to cause the output vector y_1 to track the reference input vector y_r such that the steady-state error is equal to zero. The technique consists of adding a vector comparator and integrator $z = [z_1 \ z_2]^T$ that satisfies

$$z_i(k+1) = z_i(k) + T_s (y_{r,i}(k) - y_{1,i}(k)). \tag{4.7}$$

Therefore, the open-loop system can be described by an augmented state-space representation and the controllability of the system is verified off-line. Among the most popular controller design techniques for MIMO systems, a pole placement technique is considered to impose a desired behavior of the plant in closed-loop. Therefore, the feedback gain matrix K is designed such that the eigenvalues of the closed-loop augmented system are equal to $\begin{bmatrix} 0.92 & 0.97 & 0.90 & 0.95 & 0.94 \end{bmatrix}$:

$$K = \begin{bmatrix} K_1 & | & K_2 \end{bmatrix} = 10^{-4} \times \left[\begin{pmatrix} -0.95 & -0.32 \\ -0.30 & -0.91 \end{pmatrix} \; | \; \begin{pmatrix} 21.6 & 3 & -5 \\ 2.9 & 19 & -4 \end{pmatrix} \right]; \quad (4.8)$$

$$K = \begin{bmatrix} K_1 & | & K_2 \end{bmatrix} = 10^{-4} \times \left[\begin{pmatrix} 21.6 & 3 & -5 \\ 2.9 & 19 & -4 \end{pmatrix} \; | \; \begin{pmatrix} -0.95 & -0.32 \\ -0.30 & -0.91 \end{pmatrix} \right]. \quad (4.9)$$

The control law has been implemented in C-code on a PC associated with a data acquisition board. For this system, the state variables $x(k)$ are available, and thus matrix $C = I$. Consequently, the control law is computed using the measurements. Based on a vector comparator, a simple matrix operator is computed to calculate the proportional part and also the integrator part through a recurrent algorithm under an anti-windup scheme. This integrator module with anti-windup scheme is jammed to a constant value when an actuator saturates and eliminates some possible instability problems.

Results and Comments

Step responses with respect to set-point changes are considered to validate the tracking control. Reference inputs y_r are step changes for ℓ_1 and ℓ_2 which excite the nonlinear system around the corresponding operating condition. The dynamic behavior of the levels shows that the controller is synthesized correctly (Fig. 4.3). According to the developed MIMO control law, level ℓ_i is affected by step change on level ℓ_j as presented in Fig. 4.3. Moreover, it should be noticed that the measurement noise level is very low for this application. Figure 4.4 shows the corresponding control inputs for step changes in the reference inputs. As indicated previously, ℓ_1 impacts the dynamic of ℓ_2 when the reference signal changes, consequently the corresponding control input q_1 is also affected.

4.3.3 Fault Detection and Isolation with Magnitude Estimation

In most conventional control systems, controllers are designed for the fault-free case without taking into account the possibility of fault occurrence. Due to abnormal operation or material aging, actuator or sensor faults occur in systems. First, a constant offset of -0.03 m on level sensor ℓ_1 has been created and added at instant 1000 s. In other words, the faulty measurement used by the controller is equal to $\ell_1 - 0.03$. As illustrated in Fig. 4.5, the control law tries to cancel the steady-state error created by the faulty measurement. Consequently, the real output is different from the reference input and the control law is different from its nominal value (Fig. 4.6).

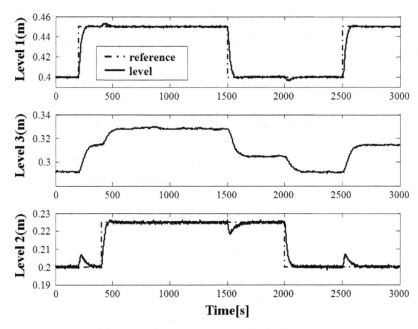

Fig. 4.3. System outputs in fault-free case

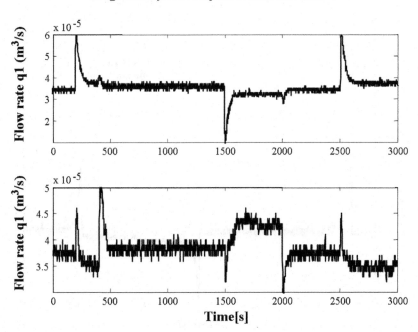

Fig. 4.4. System inputs in fault-free case

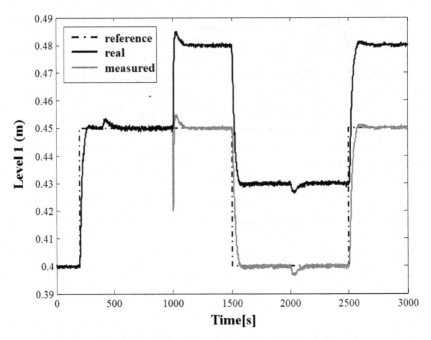

Fig. 4.5. Measured and real ℓ_1 with a bias on sensor ℓ_1

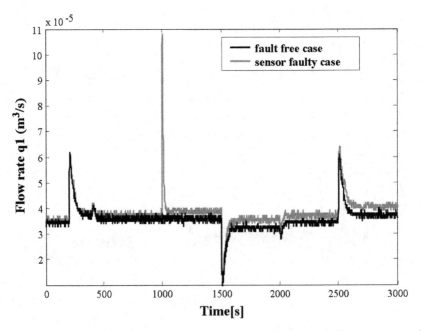

Fig. 4.6. Input flow rate in fault-free case and in the presence of a bias on sensor ℓ_1

Similarly, an actuator fault on pump 1 has been applied. A gain degradation of 80% appears abruptly at instant 1000 s. Practically, the control input applied to the system corresponds to the control input computed by the controller multiplied by a constant equal to 0.2. Since an actuator fault acts on the system as a perturbation, the system output ℓ_1 returns to its nominal value (see Fig. 4.7). With this controller, the dynamic behavior of level ℓ_2 is also affected by this fault as illustrated in Fig. 4.8.

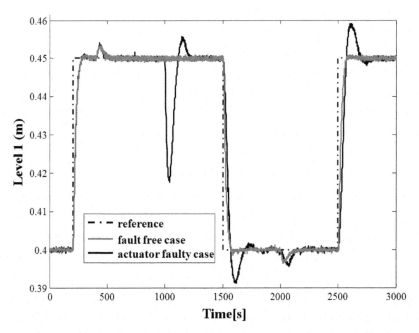

Fig. 4.7. Measured ℓ_1 in fault-free and in the presence of a bias on pump q_1

In the presence of sensor or actuator faults, system (4.4) can be represented by the discrete state-space representation

$$\begin{cases} x(k+1) = Ax(k) + Bu(k) + F_x f(k) \\ y(k) = x(k) + F_y f(k) \end{cases}, \tag{4.10}$$

where $f \in \Re^{(3+2)}$ is a common representation of sensor $f_s \in \Re^3$ and actuator $f_a \in \Re^2$ faults vectors. F_x and F_y are respectively the state and output faults matrices with $F_x = \begin{bmatrix} B & 0_{3\times3} \end{bmatrix}$ and $F_y = \begin{bmatrix} 0_{3\times2} & I_{3\times3} \end{bmatrix}$.

To detect and isolate faults, a structured residuals scheme sensitive to certain faults and insensitive to others is designed. For actuator or sensor fault representations, a unique state-space model can be established to describe the faulty system as follows (as presented in Sect. 2.5.1):

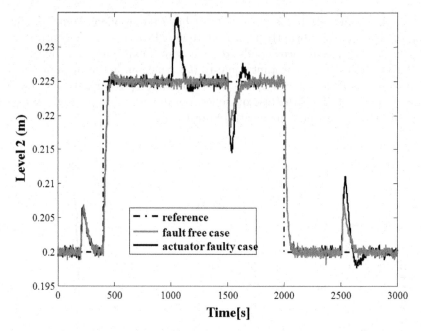

Fig. 4.8. Measured ℓ_2 in fault-free and in the presence of a bias on pump q_1

$$\begin{cases} x(k+1) = Ax(k) + Bu(k) + F_d f_d(k) + F_x^* f^*(k) \\ \quad\quad y(k) = Cx(k) + F_y^* f^*(k) \end{cases}. \tag{4.11}$$

In both cases $f_d(k)$ is the faulty unknown input vector. This unique system representation is considered for the FDI problem. While a single residual is sufficient to detect a fault, a set of structured residuals is required for fault isolation. A residual generation using unknown input observer scheme is considered in order to be sensitive to $f^*(k)$ and insensitive to fault vector $f_d(k)$ as

$$\begin{cases} w(k+1) = Ew(k) + TBu(k) + Ky(k) \\ \quad\quad \widehat{x}(k) = w(k) + Hy(k) \end{cases}, \tag{4.12}$$

where $\widehat{x}$ is the estimated state vector and w is the state of this full-order observer. E, T, K, and H are matrices to be designed to achieve unknown input decoupling requirements.

The design of the unknown input observer is achieved by solving the following equations:

$$(HC - I)F_d = 0, \tag{4.13}$$

$$T = I - HC, \tag{4.14}$$

$$E = A - HCA - K_1 C, \tag{4.15}$$

$$K_2 = EH, \tag{4.16}$$

and

$$K = K_1 + K_2. \tag{4.17}$$

Before implementing the unknown input observer, the necessary and sufficient conditions should be checked:

- $\text{Rank}(CF_d) = \text{rank}(F_d)$
- (C, A_1) is a detectable pair, where $A_1 = E + K_1 C$

As indicated below, the function, implemented in MATLAB®, synthesizes each decoupled observer according to the fault matrix F_d:

```
function [E,T,K,H]=uio_linear(A,B,C,Fd)
% Reference
% Chapter 3 - Robust residual generation via UIOs
% page 77.
% Robust Model-Based Fault Diagnosis for Dynamic Syst.
% Jie Chen and R.J. Patton
% Kluwer Avademic Publishers
% 1999
% Algorithm
% dx(t)/dt = A x(t)+ B u(t) + Fd d(t)
% y(t)     = C x(t)
% to built an UIO
% 1 a) The number of ouputs (row of C) must be greater
% than the number of unknown inputs (Column of Fd)
% 1 b) Check the rank condition for Fd and CFd
% 2 ) Compute H, T, and A1
% H  = Fd * inv[(C Fd)'* (C Fd)]*(C Fd)'
% T  = I - H C
% A1 = T A
% 3 ) Check the observability :
% If (C, A1) observable, a UIO exits and K1 can be
% computed using pole placement
% Remark : The choice of  pole placement is fixed here
% with 0.9 * eigen_value of A1
% 4 ) Compute E, K to built the following UIO
%
% dz(t)/dt = E z(t) + T B u(t) + K y(t)
% x_est(t) = z(t) + H y(t)
%
% with
%          E = A1 - K1 C
%          K = K1 + E H
```

```
% 0 ) Check input conditions
if nargin~=4,
error('Number_of_input_arguments_incorrect!...
type_help_uio_chen'),return
end


% 1 a ) The number of ouputs (row of C) must be greater
% than the number of unknown inputs (Column of Fd)
nb_Fd=size(Fd);nb_C=size(C); nb_row_C=nb_C(1);
nb_column_Fd=nb_Fd(2);
if (nb_column_Fd > nb_row_C ),
error('The_number_of_ouputs_(row_of_C)_must_be...
greater_than_the_number_of_unknown_inputs...
(column_of_Fd)'),return
end


% 1 b ) Check the rank condition for Fd and CFd
if (rank(C*Fd) ~= rank(Fd)),
error('rank(C*Fd)==rank(Fd)'),return,end


% 2 ) Compute H, T, and A1
nb_A=size(A); H=Fd*inv((C*Fd)'*(C*Fd))*(C*Fd)';
T=eye(nb_A(1))-(H*C); A1=T*A;


% 3 ) Check the observability : If (C, A1) observable,
% a UIO exits and K1 can be computed using pole
% placement
if (rank(obsv(A1,C)) ~= nb_A(1)),
error('(C,A1)_should_be_observable'),
return,
end


pole=eig(A1); K1=place(A1',C',[0.9*pole]); K1=K1';


% 4 ) Compute E, K to built the following UIO
E=A1-K1*C; K=K1+E*H;
```

For fault isolation, a bank of $(3 + 2)$ unknown input observers is constructed. Each residual vector $r_j(k) = y(k) - C\hat{x}(k)$ produced by the jth unknown input observer is used to detect a fault according to a statistical test.

In the presence of a fault, the design of structured residuals generates a residual insensitive to an expected fault. The value of the residual insensitive to the fault is close to zero unlike the other residuals. The FDI and estimation module based on the measured input and output vectors has been illustrated in the presence of a fault as previously defined. For the fault on pump 1 (respectively on sensor ℓ_1), the residuals designed for this kind of faults are close to zero while the other residuals are commonly different from zero at the time the actuator fault (respectively sensor fault) occurs. These features correspond to the expected results as illustrated in Figs. 4.9 and Fig. 4.10. Among the components, the faulty one has been correctly detected with a time delay of 16 s for the sensor and of 13 s for the actuator. In both cases and for this fault magnitude, a time delay around 14 s should be considered with a classical Page-Hinkley test for residuals evaluation and an elementary logic decision.

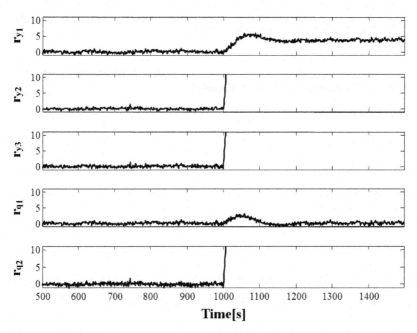

Fig. 4.9. Residuals behavior for actuator fault on pump 1

The fault magnitude estimation of the corrupted element $f_d(k)$ is extracted directly from the jth unknown input observer which is built to be insensitive to the jth fault. The state estimation is generated by this unknown input observer and matrices computations using singular value decomposition following the instructions below.

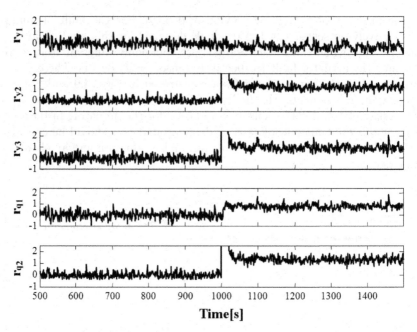

Fig. 4.10. Residuals behavior for level ℓ_1 sensor fault

```
%For  instance  for  a  fault  on  Pump  1
Fd=Bd(:,1);

%produce  a  complete  singular  value  decomposition
[T,R,M]=svd(Fd);
Abar=inv(T)*Ad*T;  A11bar=Abar(1,1);A12bar=Abar(1,2:3);
Bbar=inv(T)*Bd;B1bar=[Bbar(1,:)];

Mat_associated_to_x1=M*inv(R(1,1));
Mat_associated_to_A11=-M*inv(R(1,1))*A11bar;
Mat_associated_to_A12=-M*inv(R(1,1))*A12bar;
Mat_associated_to_B1=-M*inv(R(1,1))*B1bar;
```

Thus, the sensor input or actuator fault magnitude can be estimated as follows:

$$\widehat{f}_d(k) = V R^{-1}(\overline{\widehat{x}}_1(k+1) - \overline{A}_{11}\overline{\widehat{x}}_1(k) - \overline{A}_{12}\overline{\widehat{x}}_2(k) - \overline{B}_1 u(k)). \qquad (4.18)$$

It can be noted that the sensor fault estimation $f_d(k)$ is the last component of the estimated augmented state vector $\hat{\overline{x}}(k)$ and does not require the previous calculation.

Based on the experimental data set given as an example, the sensor (respectively actuator) fault magnitude and their estimations are illustrated in Fig. 4.11 (respectively Fig. 4.12). The fault estimation is close to zero in the fault-free case. However, the fault estimation is close to the fault magnitude when the fault occurs.

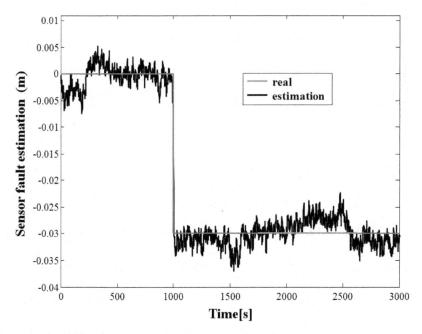

Fig. 4.11. Sensor fault magnitude and its estimation

4.3.4 Fault Accommodation

Sensor Fault Masking

In the presence of the sensor fault, only the first residual is insensitive to the fault on level sensor ℓ_1. The fault which occurs at $t = 1000$ s is isolated at instant 1016 s. Based on the sensor fault masking principle, the control law switches from the measurement to its estimation. In Fig. 4.13, it can be noted that using the fault accommodation method the real level follows the set-point unlike the case without fault accommodation. Figure 4.14 shows a zoom of Fig. 4.13 around instant 1000 s and indicates that the fault accommodation approach preserves the dynamical behavior of the system in the presence of a fault. The unknown input observer insensitive to the sensor fault occurrence provides an accurate estimate of the system output after a sensor fault occurs. The corresponding flow rates are presented in Fig. 4.15.

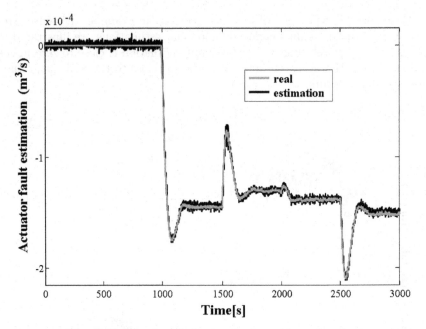

Fig. 4.12. Actuator fault magnitude and its estimation

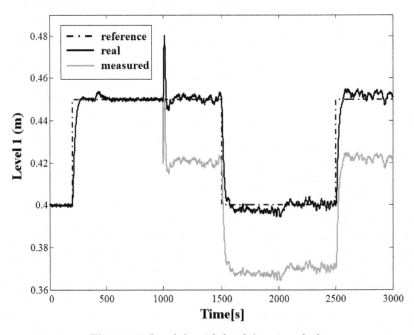

Fig. 4.13. Level ℓ_1 with level ℓ_1 sensor fault

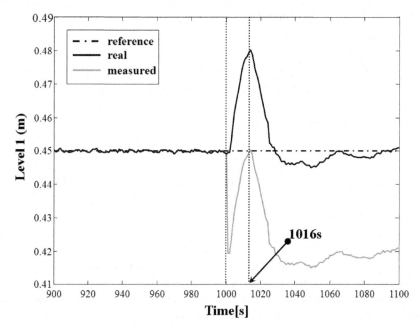

Fig. 4.14. Level ℓ_1 close to the fault occurrence

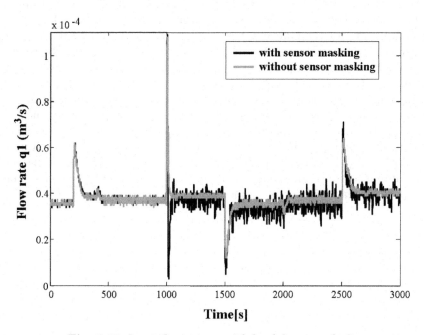

Fig. 4.15. Input flow rate q_1 with level ℓ_1 sensor fault

Actuator Fault Compensation

As presented in Chap. 2, an FTC method is used to eliminate the actuator fault effect on the system. Its goal is to compute an additional control law able to compensate for the fault effect on the system using an on-line magnitude estimation of the fault. When the fault is detected and isolated, an additional control term is computed and added to the nominal one. The new control law applied to the system is then given by

$$U(k) = (u_{nom}(k) + u_{add}(k)) + U_0. \tag{4.19}$$

If a fault is detected and isolated on the jth actuator, according to the fault magnitude estimation $\widehat{f}_d(k)$ described in the previous section, the additional control law $u_{add}(k)$ is computed on-line as

$$u_{add}(k) = -B_{add}\widehat{f}_d(k), \tag{4.20}$$

where B_{add} is computed off-line. It is equal to B^+B_j where B^+ is the pseudo-inverse of matrix B and B_j is the jth column of B.

Once the fault is isolated and estimated, the compensation control law is computed in order to reduce the fault effect on the system. With the fault accommodation method, the outputs decrease less than in the case of a classical control law as illustrated in Figs. 4.16 and 4.17. They reach the nominal values quicker because the fault is estimated and the new control law is able to compensate for the fault effect at instant 1013 s. According to Fig. 4.18, it can easily be seen that, after the fault occurrence, the time response and the overshoot of the compensated outputs are smaller than those of the faulty outputs with a classical control law. These results can be confirmed by the examination of the control input flow rate (Fig. 4.19). Without fault accommodation, the control input increases slowly trying to compensate for the fault effect. With the accommodation approach, the control input increases quicker and enables the rapid fault compensation on the controlled system outputs.

With the fault accommodation method, the dynamic behavior of the outputs after a sensor or an actuator fault occurrence is close to the nominal one compared to that without accommodation.

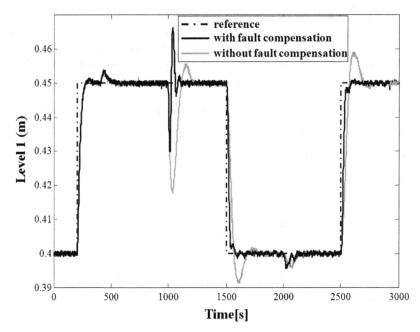

Fig. 4.16. Level ℓ_1 with pump 1 actuator fault

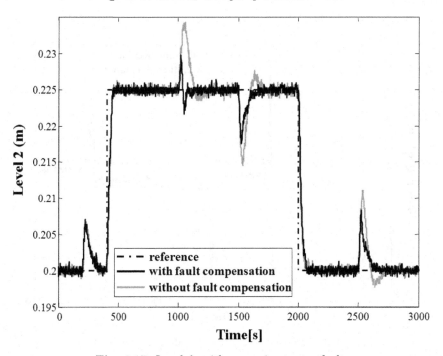

Fig. 4.17. Level ℓ_2 with pump 1 actuator fault

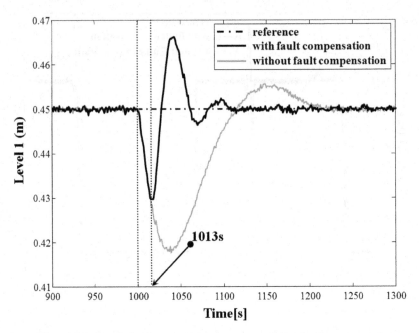

Fig. 4.18. Level ℓ_1 close to the occurrence of the fault on pump 1

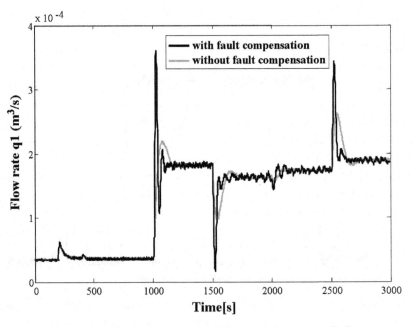

Fig. 4.19. Input flow rate q_1 with pump 1 actuator fault

4.4 Nonlinear Case

In order to extend the proposed active FTC approach based on the FDI results to the whole range of operating conditions, the nonlinear model is considered rather than a multiple model approach.

4.4.1 Nonlinear Representation

According to (4.1), the system can be written in the nonlinear affine state-space representation

$$
\begin{cases}
\dot{x}(t) = f(x(t)) + \displaystyle\sum_{i=1}^{2} g_i(x(t))u_i(t) \\
y(t) = h(x(t)) = x(t)
\end{cases}
, \tag{4.21}
$$

where the output vector $y = [\ell_1 \quad \ell_2 \quad \ell_3]^T$ is equal to the state vector x and

$$
f(x(t)) = \frac{1}{S} \begin{bmatrix} -q_{13}(t) \\ q_{32}(t) - q_{20}(t) \\ q_{13}(t) - q_{32}(t) \end{bmatrix}, \qquad g_1(x(t)) = \begin{bmatrix} \frac{1}{S} & 0 & 0 \end{bmatrix}^T,
$$

and $\qquad g_2(x(t)) = \begin{bmatrix} 0 & \frac{1}{S} & 0 \end{bmatrix}^T.$

4.4.2 Closed-loop Fault-free Case

Design and Gain Synthesis

According to (4.21), a nonlinear control law is designed to track the reference vector $y_r = [\ell_{1r} \quad \ell_{2r}]^T$. To perform this task, an input-output linearization and input-output decoupling law via a static state-feedback [41,73,98] is utilized. The system has two outputs ($p = 2$), so there are two relative degrees ρ_1 and ρ_2 to be found as defined in Chap. 2 as follows:

$$
\rho_i = \{\min l \in \aleph / \exists j \in [1,2], L_{g_j} L_f^{l-1}(x_i(t)) \neq 0\}. \tag{4.22}
$$

Applying (4.22) to (4.21), the degrees ρ_1 and ρ_2 are equal to 1. Then, the decoupling matrix $\Delta(x(t))$ is given by

$$
\Delta(x(t)) = \begin{bmatrix} L_{g_1}x_1(t) & L_{g_2}x_1(t) \\ L_{g_1}x_2(t) & L_{g_2}x_2(t) \end{bmatrix} = \begin{bmatrix} \frac{1}{S} & 0 \\ 0 & \frac{1}{S} \end{bmatrix}. \tag{4.23}
$$

According to the condition $rank\,(\Delta(x(t))) = 2$, the three-tank system can be statically decoupled. The control law synthesis requires a vector Δ_0 defined as

$$
\Delta_0(x) = \begin{bmatrix} L_f x_1(t) \\ L_f x_2(t) \end{bmatrix} = \begin{bmatrix} -\frac{1}{S}q_{13}(x) \\ \frac{1}{S}(q_{32}(x) - q_{20}(x)) \end{bmatrix}, \tag{4.24}
$$

where each inter flow rate depends on the levels.

A linearizing static feedback $u(t) = -\Delta^{-1}(x)\Delta_0(x) + (\Delta^{-1}(x))v(t)$ is then designed for the three tank system as follows:

$$u(t) = \begin{bmatrix} q_1(t) \\ q_2(t) \end{bmatrix} = \begin{bmatrix} -\frac{1}{S}q_{13}(x(t)) \\ \frac{1}{S}(q_{32}(x(t)) - q_{20}(x(t))) \end{bmatrix} + \begin{bmatrix} S & 0 \\ 0 & S \end{bmatrix} v(t), \qquad (4.25)$$

where $v(t)$ represents the input vector of the equivalent linearized model.

According to the value of relative degrees ρ_1 and ρ_2, the closed-loop system can be described as two independent linear SISO subsystems equivalent to a unique integrator expressed as

$$\begin{cases} \dfrac{\ell_1(s)}{v_1(s)} = \dfrac{1}{s} \\[2mm] \dfrac{\ell_2(s)}{v_2(s)} = \dfrac{1}{s} \end{cases}, \qquad (4.26)$$

where s is the Laplace variable.

Remark 4.2. It can be noted that $\sum_{i=1}^{3} r_i = 2 < 3$. The system has an unobservable subspace of dimension one which can be associated with tank 3. Due to the stable property of this unobservable subspace, the linearizing control law can be applied to the three-tank system.

Each linear SISO subsystem (4.26) is unstable, and therefore a second control law should be designed to stabilize it and to assign the dynamic behavior following the linear control theory. A proportional output feedback $v_i(t) = K_i(y_{r,i}(t) - \ell_i(t))$ is applied to each ith decoupled subsystem which is equivalent to

$$\frac{\ell_i(s)}{y_{r,i}(s)} = \frac{K_i}{s + K_i}, \qquad (4.27)$$

where K_i represents the proportional gain and $y_{r,i}$ the reference input.

The block diagram built using Simulink® (Fig. 4.20) summarizes the designed nonlinear controller with the main blocks "Stabilization" and "Linearization." Each Simulink® block corresponding to the previous equations is presented in Figs. 4.21 and 4.22. Moreover, it can be highlighted that the C-code used for the on-line implementation of the nonlinear control law is based on the elements presented in each block based on elementary mathematical functions or operators.

The parameter estimation shows modeling errors, and corrupts the "ideal" decoupled subsystem as defined in (4.26). This problem generates a steady-state error in closed-loop. To eliminate the difference between the desired output $y_{r,i}$ and the level ℓ_i, an integrator is added to each controller of the stabilization feedback.

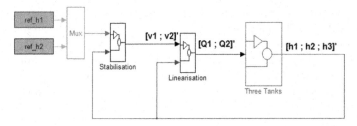

Fig. 4.20. Block diagram for nonlinear control law design in fault-free case

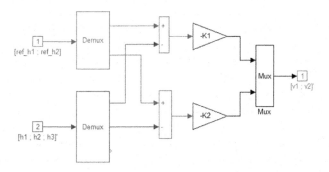

Fig. 4.21. Stabilization Simulink® block for nonlinear control law design

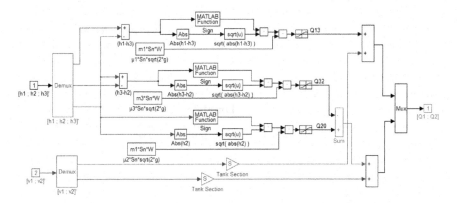

Fig. 4.22. Linearization Simulink® block for nonlinear control law design

Results and Comments

The control law has been applied to the three-tank system in the fault-free case. The results shown in the following figures are step responses obtained for different reference inputs. All experiments are illustrated for a range of 4000 s. The outputs ℓ_1 and ℓ_2 follow the reference inputs $y_{r,1}$ and $y_{r,2}$ from an initial level of 0 m until a maximum level of 0.5 m for ℓ_1 (Fig. 4.23) and 0.4m for ℓ_2 (Fig. 4.24) with different steps. Generally speaking, the variance of the noise is more important during the first 1000 samples on the control signal q_1 (Fig. 4.25). It is due to the fact that the level ℓ_1 is very low in the first tank and the output of the pump is very high: the noise corresponds to the effect of the water fall (Fig. 4.23). The same could be said about the control signal q_2 (Fig. 4.26). According to the considered closed-loop strategy, the system is completely decoupled into two subsystems, since level ℓ_2 is never affected by the variations of level ℓ_1 (which is not the case in the previously studied linear case).

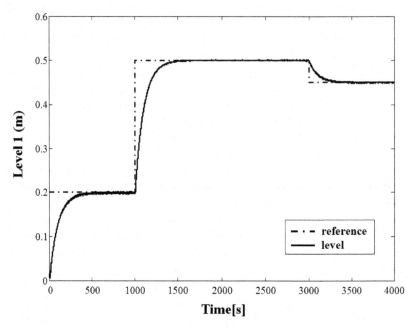

Fig. 4.23. Level ℓ_1 in fault-free case

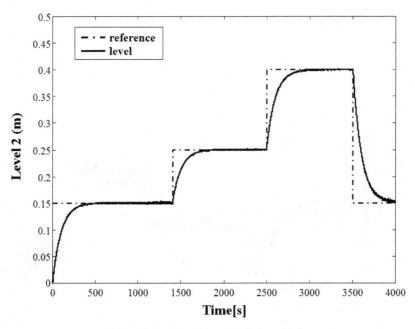

Fig. 4.24. Level ℓ_2 in fault-free case

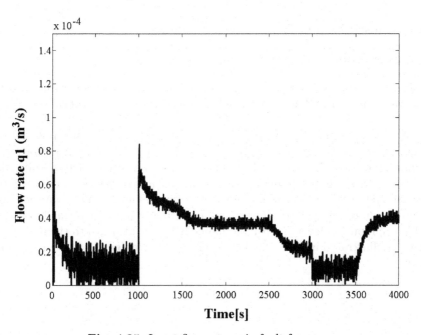

Fig. 4.25. Input flow rate q_1 in fault-free case

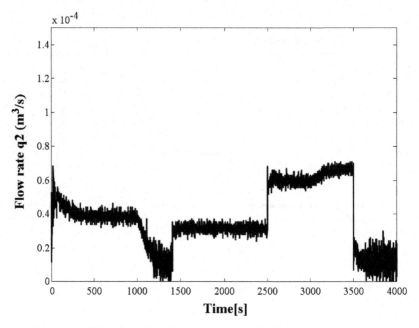

Fig. 4.26. Input flow rate q_2 in fault-free case

4.4.3 Closed-loop in the Presence of Faults

Actuator Fault

Unlike the previous study, which was carried out around an operating point, actuator and sensor faults will now be considered for the whole operating range.

A first experiment consists of assuming a degradation of pump 1 of 80% to occur at instant $t = 2000$ s. As presented in Fig. 4.27, the output ℓ_1 does not follow its reference input $y_{r,1}$. However, it can also be noticed that, despite the degradation of pump 1 (see Fig. 4.28), the system is perfectly decoupled into two subsystems. The control law associated with pump 2 compensates for the degradation of pump 1 considered as a disturbance in order to maintain the reference input on level ℓ_2 with a dynamic behavior identical to the fault-free case. As illustrated in Fig. 4.29 only level ℓ_3 is affected by the presence of this fault.

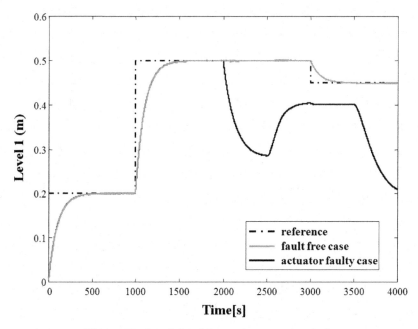

Fig. 4.27. Level ℓ_1 with pump 1 actuator fault

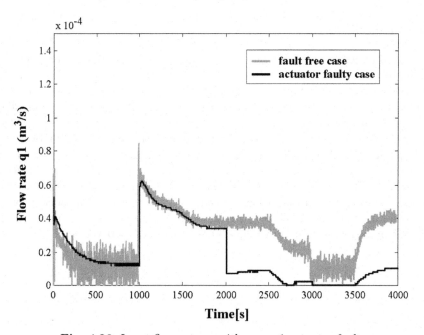

Fig. 4.28. Input flow rate q_1 with pump 1 actuator fault

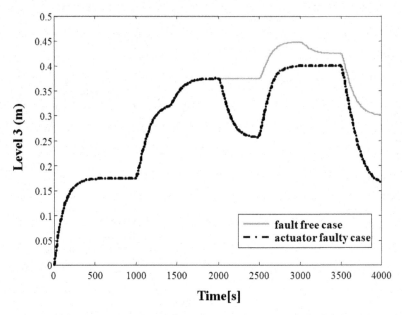

Fig. 4.29. Level ℓ_3 with pump 1 actuator fault

Sensor Fault

A second experiment corresponds to an incipient degradation on level sensor ℓ_1 with a gain equal to $5 \times 10^{-5}\ ms^{-1}$. This fault is assumed to occur at $t = 2000\ s$ as presented in Fig. 4.30. This type of fault often occurs on sensors and is generally due to material aging. When the fault appears, the controller brings the faulty measurement ℓ_1 back to the corresponding reference value $y_{r,1}$. The real level is far from the desired value (Fig. 4.30) and moreover the subsystem associated with ℓ_1 is affected by the dynamic behavior of output ℓ_2. Indeed, the nonlinear model considered in the decoupled control law synthesis does not correspond to the actual system. Consequently, the robustness of the closed-loop against disturbances is affected by the sensor fault occurrence. Indeed, the "measured" level ℓ_1 (affected by the sensor fault) cannot reach the set-point as illustrated in Fig. 4.30 after instant 2500 s.

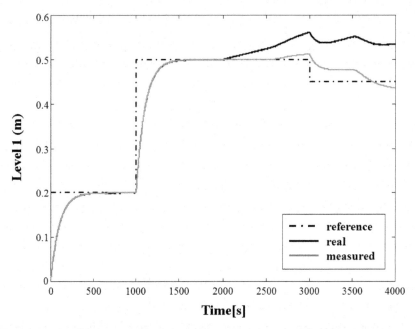

Fig. 4.30. Level ℓ_1 with a drift on sensor level ℓ_1

4.4.4 Sensor Fault-tolerant Control Design

Sensor Fault Detection and Isolation with Magnitude Estimation

In the presence of faults on each measured level, the three-tank system defined in (4.1) can be described by the following state-space representation:

$$
\begin{cases}
\begin{bmatrix} \dot{l}_1(t) \\ \dot{l}_2(t) \\ \dot{l}_3(t) \end{bmatrix} = \begin{bmatrix} \dfrac{-q_{13}(l_1,l_3)}{S} \\[2mm] \dfrac{q_{32}(l_2,l_3)-q_{20}(l_2)}{S} \\[2mm] \dfrac{q_{13}(l_1,l_3)-q_{32}(l_2,l_3)}{S} \end{bmatrix} + \begin{bmatrix} \frac{1}{S} & 0 \\ 0 & \frac{1}{S} \\ 0 & 0 \end{bmatrix} \begin{bmatrix} u_1(t) \\ u_2(t) \end{bmatrix} \\[8mm]
\begin{bmatrix} y_1(t) \\ y_2(t) \\ y_3(t) \end{bmatrix} = \begin{bmatrix} l_1(t) \\ l_2(t) \\ l_3(t) \end{bmatrix} + \begin{bmatrix} f_{s_1}(t) \\ f_{s_2}(t) \\ f_{s_3}(t) \end{bmatrix}
\end{cases}
\tag{4.28}
$$

with

$$
\begin{cases}
q_{13}(t) = c_1 \sqrt{l_1(t) - l_3(t)} \\[2mm]
q_{20}(t) = c_2 \sqrt{l_2(t)} \\[2mm]
q_{13}(t) = c_3 \sqrt{l_3(t) - l_2(t)}
\end{cases}
\tag{4.29}
$$

where parameters c_1, c_2, and c_3 are constant and can be calculated based on constant parameters presented in Table 4.1 according to (4.2) and (4.3).

Under sensor faults assumptions, as proposed in Sect. 2.5.1, the previous state-space representation can be defined through the following augmented state:

$$x(t) = \begin{bmatrix} l_1(t) \\ l_2(t) \\ l_3(t) \\ f_{s_1}(t) \\ f_{s_2}(t) \\ f_{s_3}(t) \end{bmatrix}, \tag{4.30}$$

where sensor faults f_{s_i} are considered as a linear system excited by external inputs $\bar{f}_{s_i}$:

$$\begin{cases} \dot{f}_{s_1}(t) = \gamma_1 f_{s_1}(t) + \bar{f}_{s_1}(t) \\ \dot{f}_{s_2}(t) = \gamma_2 f_{s_2}(t) + \bar{f}_{s_2}(t) \\ \dot{f}_{s_3}(t) = \gamma_3 f_{s_3}(t) + \bar{f}_{s_3}(t) \end{cases}. \tag{4.31}$$

The system can be expressed as a *"pseudo-actuator"* faults form as follows:

$$\dot{x}(t) = \underbrace{\begin{bmatrix} -c_1\sqrt{l_1(t)-l_3(t)} \\ c_3\sqrt{l_3(t)-l_2(t)} - c_2\sqrt{l_2(t)} \\ c_1\sqrt{l_1(t)-l_3(t)} - c_3\sqrt{l_3(t)-l_2(t)} \\ \gamma_1 f_{s_1}(t) \\ \gamma_2 f_{s_2}(t) \\ \gamma_3 f_{s_3}(t) \end{bmatrix}}_{f(x(t))} + \underbrace{\begin{bmatrix} \frac{1}{S} & 0 \\ 0 & \frac{1}{S} \\ 0 & 0 \\ 0 & 0 \\ 0 & 0 \\ 0 & 0 \end{bmatrix}}_{g(x(t))} \begin{bmatrix} u_1(t) \\ u_2(t) \end{bmatrix}$$

$$+ \underbrace{\begin{bmatrix} 0 & 0 & 0 \\ 0 & 0 & 0 \\ 0 & 0 & 0 \\ 1 & 0 & 0 \\ 0 & 1 & 0 \\ 0 & 0 & 1 \end{bmatrix}}_{F(t)} \underbrace{\begin{bmatrix} \bar{f}_{s_1}(t) \\ \bar{f}_{s_2}(t) \\ \bar{f}_{s_3}(t) \end{bmatrix}}_{f(t)}. \tag{4.32}$$

To be sensitive to some faults and insensitive to others, the fault vector is broken down into two parts as follows:

$$\dot{x}(t) = f(x(t)) + g(x(t))u(t) + \underbrace{\begin{bmatrix} 0 \\ 0 \\ 0 \\ 1 \\ 0 \\ 0 \end{bmatrix}}_{F_1(t)} \bar{f}_{s_1}(t) + \underbrace{\begin{bmatrix} 0 & 0 \\ 0 & 0 \\ 0 & 0 \\ 0 & 0 \\ 1 & 0 \\ 0 & 1 \end{bmatrix}}_{D_1(t)} \underbrace{\begin{bmatrix} \bar{f}_{s_2}(t) \\ \bar{f}_{s_3}(t) \end{bmatrix}}_{d_1(t)}. \tag{4.33}$$

Based on the previous state-space, an associated detectability index is computed as defined in (2.80). For the first sensor (*i.e.* $i = 1$), ρ_1 is defined as

$$\rho_1 = \min\{\zeta \in \mathbb{N} | L_{F_1} L_f^{\zeta-1} \hat{x}_1(t) \neq 0\}, \tag{4.34}$$

where $\hat{x}_1(t)$ is the estimation of the first component of $x(t)$.

For the considered system, ρ_1 is equal to one: only output y_1 is affected by f_{s_1}. Thus, the state-space representation insensitive to fault f_{s_1} is defined as

$$\tilde{x}(t) = \phi_{f_{s_1}}(x(t), u(t)) = \begin{bmatrix} \tilde{x}_a(t) \\ \tilde{x}_b(t) \end{bmatrix} = \begin{bmatrix} y_1(t) \\ \phi_1(x(t), u(t)) \end{bmatrix}, \tag{4.35}$$

where $\tilde{x}_a(t) = y_1(t) = l_1(t) + f_{s_1}(t)$ and $\tilde{x}_b(t) = [l_1(t) \quad l_2(t) \quad l_3(t) \quad f_{s_2}(t) \quad f_{s_3}(t)]^T$. It can be noted that $\phi_1(x(t), u(t))$ is independent of $f_{s_1}(t)$.

Therefore, the faulty sensor nonlinear affine state-space representation is described through the following decoupled form:

$$\begin{cases} \dot{\tilde{x}}_a(t) = \dot{l}_1(t) + \dot{f}_{s_1}(t) = \left(-c_1\sqrt{l_1(t) - l_3(t)}\right) + \dfrac{1}{S}u_1(t) + \dot{f}_{s_1}(t) = \dot{y}_1(t) \\[2mm]
\dot{\tilde{x}}_b(t) = \begin{bmatrix} \dot{l}_1(t) \\ \dot{l}_2(t) \\ \dot{l}_3(t) \\ \dot{f}_{s_2}(t) \\ \dot{f}_{s_3}(t) \end{bmatrix} = \begin{bmatrix} -c_1\sqrt{l_1(t) - l_3(t)} \\ c_3\sqrt{l_3(t) - l_2(t)} - c_2\sqrt{l_2(t)} \\ c_1\sqrt{l_1(t) - l_3(t)} - c_3\sqrt{l_3(t) - l_2(t)} \\ \gamma_2 f_{s_2}(t) \\ \gamma_3 f_{s_3}(t) \end{bmatrix} \\[2mm]
\qquad\qquad + \begin{bmatrix} \frac{1}{S} & 0 \\ 0 & \frac{1}{S} \\ 0 & 0 \\ 0 & 0 \\ 0 & 0 \end{bmatrix} \begin{bmatrix} u_1(t) \\ u_2(t) \end{bmatrix} + \begin{bmatrix} 0 & 0 \\ 0 & 0 \\ 0 & 0 \\ 1 & 0 \\ 0 & 1 \end{bmatrix} \begin{bmatrix} \bar{f}_{s_2}(t) \\ \bar{f}_{s_3}(t) \end{bmatrix} \\[2mm]
\tilde{y}_a(t) = y_1(t) \\[2mm]
\tilde{y}_b(t) = \begin{bmatrix} l_2(t) + f_{s_2}(t) \\ l_3(t) + f_{s_3}(t) \end{bmatrix} \end{cases}$$

$$\tag{4.36}$$

The subsystem insensitive to f_{s_1} defined in (4.37) is considered in order to generate a residual r_1 of zero mean in both fault-free case and faulty case:

$$
\begin{cases}
\dot{\tilde{x}}_b(t) = \begin{bmatrix} \dot{\tilde{x}}_{b_1}(t) \\ \dot{\tilde{x}}_{b_2}(t) \\ \dot{\tilde{x}}_{b_3}(t) \\ \dot{\tilde{x}}_{b_4}(t) \\ \dot{\tilde{x}}_{b_5}(t) \end{bmatrix} = \underbrace{\begin{bmatrix} -c_1\sqrt{\tilde{x}_{b_1}(t) - \tilde{x}_{b_3}(t)} \\ c_3\sqrt{\tilde{x}_{b_3}(t) - \tilde{x}_{b_2}(t)} - c_2\sqrt{\tilde{x}_{b_2}(t)} \\ c_1\sqrt{\tilde{x}_{b_1}(t) - \tilde{x}_{b_3}(t)} - c_3\sqrt{\tilde{x}_{b_3}(t) - \tilde{x}_{b_2}(t)} \\ \gamma_2\tilde{x}_{b_4}(t) \\ \gamma_3\tilde{x}_{b_5}(t) \end{bmatrix}}_{\tilde{f}(\tilde{x}_b(t))} \\[2em]
\qquad + \underbrace{\begin{bmatrix} \frac{1}{S} & 0 \\ 0 & \frac{1}{S} \\ 0 & 0 \\ 0 & 0 \\ 0 & 0 \end{bmatrix}}_{\tilde{g}(\tilde{x}_b(t))} \begin{bmatrix} u_1(t) \\ u_2(t) \end{bmatrix} + \underbrace{\begin{bmatrix} 0 & 0 \\ 0 & 0 \\ 0 & 0 \\ 0 & 0 \\ 1 & 0 \\ 0 & 1 \end{bmatrix}}_{\tilde{D}_1} \begin{bmatrix} \bar{f}_{s_2}(t) \\ \bar{f}_{s_3}(t) \end{bmatrix} \\[2em]
\tilde{y}_b(t) = \tilde{y}_1(t) = \underbrace{\begin{bmatrix} \tilde{x}_{b_2}(t) + \tilde{x}_{b_4}(t) \\ \tilde{x}_{b_3}(t) + \tilde{x}_{b_5}(t) \end{bmatrix}}_{\tilde{h}_1(\tilde{x}_b(t))}
\end{cases} \quad (4.37)
$$

Indeed, based on the previous reduced system, represented by the following form:

$$
\begin{cases}
\dot{\tilde{x}}_b(t) = \tilde{f}(\tilde{x}_b(t)) + \tilde{g}(\tilde{x}_b(t))u(t) + \tilde{D}_1 \begin{bmatrix} \bar{f}_{s_2}(t) \\ \bar{f}_{s_3}(t) \end{bmatrix} . \\
\tilde{y}_b(t) = \tilde{y}_1(t) = \tilde{h}_1(\tilde{x}_b(t))
\end{cases} \quad (4.38)
$$

A classical extended Luenberger observer should be considered to estimate the state as follows:

$$
\dot{\hat{\tilde{x}}}_b(t) = \tilde{f}(\hat{\tilde{x}}_b(t)) + \tilde{g}(\hat{\tilde{x}}_b(t))u(t) + L_1\left(\tilde{y}_b(t) - \hat{\tilde{y}}_b(t)\right), \quad (4.39)
$$

where L_1 corresponds to the observer gain computed at each step time so that the eigenvalues of $\left(\frac{\partial \tilde{f}(\tilde{x}_b(t))}{\partial \tilde{x}_b(t)} - L_1 \frac{\partial \tilde{h}(\tilde{x}_b(t))}{\partial \tilde{x}_b(t)}\right)$ are stable and follow the dynamic specifications.

The same calculation is applied to each sensor fault. Therefore, a bank of three classical extended Luenberger observers, where each of them is decoupled to one single fault, is used to detect and isolate sensor faults.

In the fault-free case, the residuals generated by the three extended Luenberger observers are close to "zero." Each residual is affected by estimation errors issued from some modeling errors as illustrated in Fig. 4.31. With an appropriate evaluation method, no faults are isolated. However, in the presence of an incipient fault on the first sensor, Fig. 4.32 shows that at instant

2000 s, two of the three residuals present some incipient variation. It can be noted that the residuals have the same evolution as the fault. As said previously, only the first residual is insensitive to the fault on level sensor ℓ_1. The residual evaluation detects this evolution at instant 2009 s.

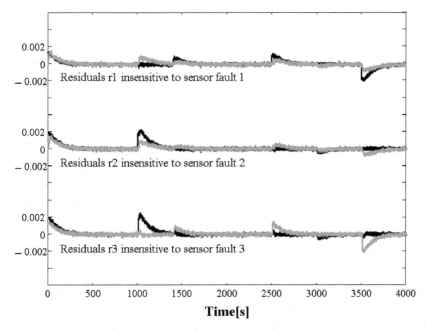

Fig. 4.31. Dynamic behavior of residuals in fault-free case

Sensor Masking Approach

As presented in the previous experiment, the observer insensitive to a sensor fault allows one to compute an estimation of the state $\hat{x}$ which is not affected by the fault. Consequently, an estimation of the output corrupted by the fault is considered in the control law rather than the measured (and corrupted) one. In the presence of sensor faults, the faulty measurements directly affect the closed-loop behavior or the state estimator. Moreover, the controller aims at canceling the error between the measurement and its reference input. However, the real output is far from the desired value and may drive the system to its physical limitations or even to instability as presented in Fig. 4.30. Sensor FTC can be obtained by computing a new control law using a fault-free estimation of the faulty element to prevent faults from developing into failures and to minimize the fault effects on the system performance and safety. From the control point of view, sensor FTC does not require any modification of the control law and is also called "sensor masking" as suggested by [131]. The

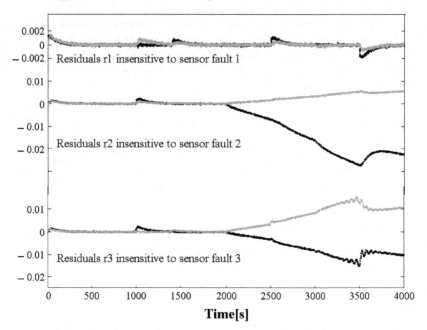

Fig. 4.32. Dynamic behavior of residuals with a drift on sensor level ℓ_1

only requirement is that the "estimator" provides an accurate estimate of the system output after an instrument fault occurs. Then, as previously said, only the first residual is insensitive to the fault on level sensor. The fault is isolated at instant 2009 s and consequently the control law switches from measurement ℓ_1 to its estimation $\hat{\ell}_1$ issued from the extended Luenberger observer decoupled to the sensor fault on ℓ_1. With the FTC method, the real level ℓ_1 follows its reference. When the fault appears, the measured level ℓ_1 has the same evolution as the fault. Figure 4.33 shows the abilities of the sensor FTC method to compensate for incipient faults.

The FDI of the developed strategy is of paramount importance to compensate for these faults and to preserve system performances.

4.4.5 Actuator Fault-tolerant Control Design

Actuator FDI with Magnitude Estimation

According to the nonlinear control law synthesis, the input-output decoupled transfer function (4.26) of each subsystem $\left(\dfrac{\ell_i}{v_i}\right)$ with ($i = 1, 2$) is linear and equal to one integrator. Each of them could then be expressed as the following classical Brunovsky canonical form [65]:

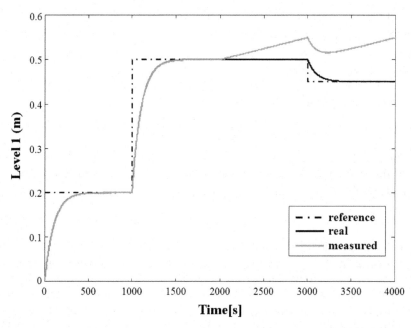

Fig. 4.33. Level ℓ_1 with sensor fault masking in the presence of a drift on sensor level ℓ_1

$$\begin{cases} \dot{\tilde{x}}(t) = A\tilde{x}(t) + Bv_i(t) \\ y_i(t) = C\tilde{x}(t) \end{cases}, \quad 1 \le i \le 2, \tag{4.40}$$

where for the three-tank system $A = [0]$, $B = C = [1]$, and $y_i(t) = \ell_i(t)$.

As shown in Figs. 4.16 and 4.18 in the linear case, based on the nonlinear decoupled control law, a fault on actuator i affects the reference input i. According to the actuator faulty linear representation presented in Section 2.3.1, the Brunovsky canonical form (4.40) can be represented as

$$\begin{cases} \dot{\tilde{x}}(t) = A\tilde{x}(t) + Bv_i(t) + Bd_i(t) \\ y_i(t) = C\tilde{x}(t) \end{cases}, \quad 1 \le i \le 2, \tag{4.41}$$

where $d_i(t)$ represents an image of the actuator fault effect on the ith decoupled subsystem ($i = 1, 2$).

Furthermore, the Brunovsky canonical form will prove useful dealing with the problem of fault magnitude estimation. Indeed, in order to detect, isolate, and estimate the fault, a state observer can be associated with each subsystem (4.26) and synthesized based on the Brunovsky canonical form through the following form:

$$\begin{cases} \dot{\hat{x}}(t) = A\hat{x}(t) + Bv_i(t)) + L(y_i(t) - C\hat{x}(t)) \\ \hat{y}_i(t) = C\hat{x}(t) \end{cases}, \quad 1 \le i \le 2, \quad (4.42)$$

where L is the observer gain, $\hat{x}$ defines the estimated state vector, and $\hat{y}_i$ represents the estimated output. The estimation error vector, denoted $\varepsilon(t) = \tilde{x}(t) - \hat{\tilde{x}}(t)$, is equivalent to

$$\dot{\varepsilon}(t) = (A - LC)\varepsilon(t) + d_i(t). \quad (4.43)$$

Due to the property of matrix C being equal to 1, the observation error is directly computed on-line with the measurements and it provides an estimation of the fault magnitude as follows:

$$\hat{d}_i(t) = \dot{\varepsilon}(t) - (A - LC)\varepsilon(t). \quad (4.44)$$

This vector should be exploited for FDI. Then the fault magnitude estimation $\hat{d}_i$ is used as a residual in order to detect and isolate the actuator fault occurrence. Under the abrupt degradation of 80% occurring at instant $t = 2000\ s$ on the pump 1, it can be seen that the experimental estimation of the fault d_1 is zero mean in the absence of fault and nonzero otherwise as illustrated in Fig. 4.34. The estimation of the fault d_2 is close to zero (Fig. 4.35): the fault (pump degradation) concerns only subsystem 1 and d_2 concerns only subsystem 2.

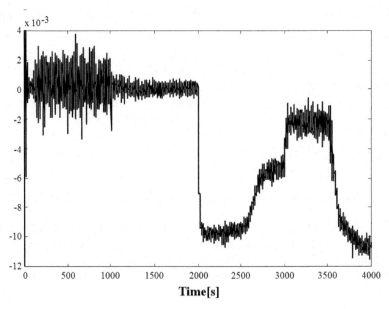

Fig. 4.34. d_1 fault magnitude estimation with an abrupt loss of effectiveness of pump 1

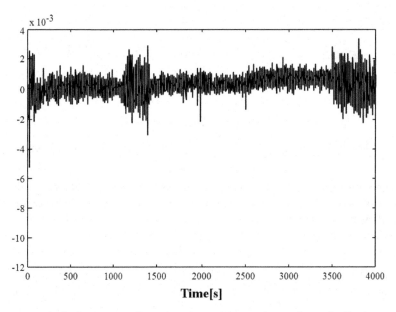

Fig. 4.35. d_2 fault magnitude estimation with an abrupt loss of effectiveness of pump 1

Indeed, a residual evaluation algorithm can be performed by the direct fault magnitude evaluation through a statistical test in order to monitor the process. A bank of residuals can be considered to generate fault signatures and to isolate faults as presented in Chap. 2 through a decision logic for instance.

Actuator Fault Accommodation

In the spirit of fault compensation principles, an additional control term v_i, added to the nominal one $v_{nom,i}$, is computed in order to eliminate the actuator fault effect. Based on the linear system, controlled by a classical state feedback, the total control law is then computed such as

$$v_i(t) = v_{nom,i}(t) + v_{add,i}. \tag{4.45}$$

Fig. 4.36 represents the implementation of the FTC system using Simulink®. When the FDI module is active, the "accommodation" block will compute an additive control law added to the nominal one.

Therefore, each decoupled closed-loop system, defined in (4.40), is given by

$$\begin{cases} \dot{\hat{x}}(t) = A\hat{x}(t) + Bv_{nom,i}(t) + Bv_{add,i}(t) + Bd_i(t) \\ \hat{y}_i(t) = C\hat{x}(t) \end{cases}, \quad 1 \leq i \leq 2. \tag{4.46}$$

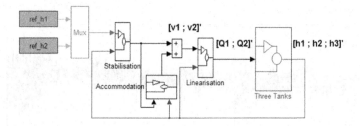

Fig. 4.36. Block diagram for actuator fault accommodation design

The additional control law is computed on-line such that the faulty system is as close as possible to the fault-free system behavior. This consists of solving the following equation:

$$Bv_{add,i}(t) + Bd_i(t) = 0. \tag{4.47}$$

Then the closed-loop system is driven by a new control law composed of the nominal control law with the additional one with the following property:

$$v_{add,i}(t) = -d_i(t). \tag{4.48}$$

Since $d_i(t)$ is unknown, its estimation $\hat{d}_i(t)$ given by (4.44) should be considered in the proposed compensation approach.

The "fault accommodation" block is shown in Fig. 4.37. As presented, each $d_i(t)$ is estimated through a classical Luenberger observer. Based on each $\hat{d}_i(t)$, the additive control law $v_{add,i}$ is generated.

Based on the on-line actuator fault magnitude estimations ($\widehat{d_1}(t)$ and $\widehat{d_2}(t)$), the proposed FTC approach is able to compensate for the fault effect when the fault appears with a small delay depending on the FDI module performance. As illustrated in Fig. 4.38, the output ℓ_1 decreases less than without fault accommodation, then it reaches its reference value.

As indicated in the proposed method, the additive control law $v_{add,1}(t)$ is able to compensate for the fault effect at the instant the fault is detected. From a practical point of view, a simple numerical filter with an appropriate gain is considered in order to reduce the noise on the fault magnitude estimation. Then the closed-loop system is driven by a new control law composed of the nominal control law with the additional one as presented in Fig. 4.39.

Nevertheless, the level ℓ_1 is sensitive to the change of the reference of ℓ_2 as presented in Fig. 4.40 around $t = 2500$s. Due to modeling errors, the system is not exactly decoupled. From a practical point of view, the FTC approach in the presence of fault preserves the dynamical behavior of the system.

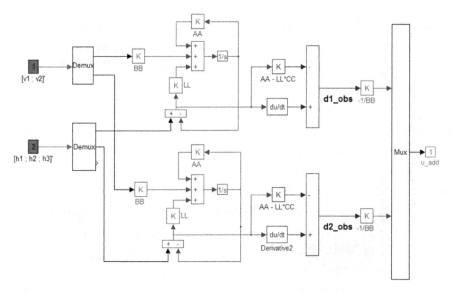

Fig. 4.37. Actuator fault accommodation design

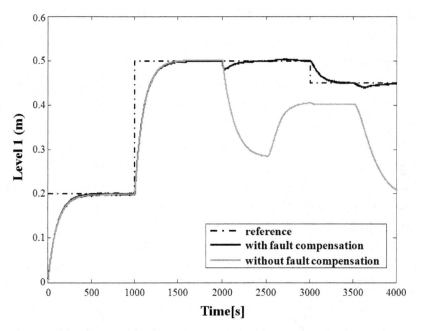

Fig. 4.38. Level ℓ_1 with or without fault accommodation method in the presence of an abrupt loss of effectiveness of pump 1

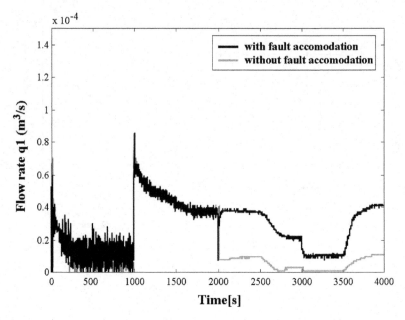

Fig. 4.39. Accommodated input flow rate q_1 with an abrupt loss of effectiveness of pump 1

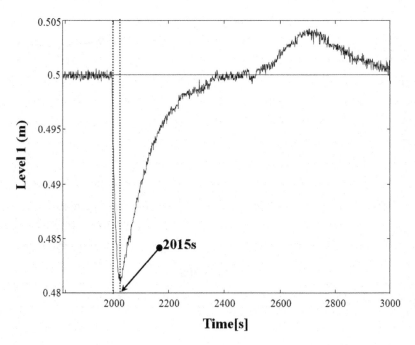

Fig. 4.40. Level ℓ_1 around $t = 2500$s with an abrupt loss of effectiveness of pump 1

4.4.6 Fault-tolerant Control Design Against Major Actuator Failures

Actuator Fault-tolerant Control Synthesis

The FTC strategy, presented in this paragraph, deals with major actuator failures as a blocking or a complete loss of an actuator appearing on systems without actuator redundancy. In the event of this kind of failure, it is impossible to maintain the faulty system to a desirable level of performance based on the FTC method developed in Chap. 2. Thus, the proposed method consists of leading the system to its optimal operating order with respect to desirable performances. Moreover, both static and dynamic performances are taken into account in the generation of the optimal reference trajectory, allowing the faulty system to be driven to its optimal operating order. The distribution of the available energy among the healthy actuators, the accessibility, and the stability conditions are also considered. This corresponds to a nonlinear constrained dynamical problem.

Consider the following m-inputs, m-outputs nonlinear system:

$$\begin{cases} \dot{x}(t) = f(x(t)) + \sum_{i=1}^{m} g_i(x(t))u_i(t) \\ Y(t) = h(x(t)) \end{cases} \tag{4.49}$$

where $x \in \Re^n$ is the state vector, $u = \begin{bmatrix} u_1 & \dots & u_m \end{bmatrix}^T \in \Re^m$ is the control vector, and $y \in \Re^m$ is the output vector to be controlled. $f(.)$, $h(.)$ and $g_i(.)$, $(i = 1, ..., m)$ are smooth vector fields. This model is defined on an open set of (Ψ) under the following physical constraints:

$$(\Psi) : \begin{cases} x_{min} \le x(t) \le x_{max} \\ u_{min} \le u(t) \le u_{max} \\ y_{min} \le y(t) \le y_{max} \end{cases} \tag{4.50}$$

An input-output linearization and decoupling law via a static state feedback is assumed to control the plant according to the set (Ψ).

When a major failure appears on the jth actuator (blocking or complete loss), the nonlinear system becomes

$$(\Sigma_j) : \begin{cases} \dot{x}_f(t) = f(x_f(t)) + \sum_{i=1}^{m} g_i(x_f(t))u_{i,f}(t) + g_j(x_f(t))d_j \\ y_f(t) = h(x_f(t)) \end{cases} \tag{4.51}$$

which is equivalent to:

$$(\Sigma_j): \begin{cases} \dot{x}_f(t) = f_j(x_f(t)) + \sum_{i=1,\ i\neq j}^{m} g_i(x_f(t))u_{i,f}(t) \\ y_f(t) = h(x_f(t)) \end{cases}, \qquad (4.52)$$

where

$$f_j(x_f(t)) = f(x_f(t)) + g_j(x_f(t))d_j, \qquad (4.53)$$

with $u_f = [u_{1,f} \quad \cdots \quad u_{j-1,f} \quad u_{j+1,f} \quad u_{m,f}]^T \in \Re^{m-1}$, $x_f \in \Re^n$, $y_f \in \Re^m$ and d_j is a constant corresponding to the jth actuator failure value.

If the actuator is blocked, d_j is equal to the blocking value; however if the actuator is lost (out of order), d_j is null. Thus, in the case of a major actuator failure, the plant dynamic structure itself changes suddenly. The nominal control no longer maintains the damaged system (Σ_j) at an admissible level of performance, regardless of the reference trajectories. The set of available inputs has decreased. Thus, only $(m-1)$ outputs could be tracked now.

The active FTC for major actuator failures, developed here, is composed of a failure detection and isolation module, the redesign of a new control law, and the computation of suitable reference inputs. FDI is realized via the measurement of inputs and assumed without false alarms and missed detection problems. Considering a major failure, it is impossible to preserve an acceptable level of performance, regardless of the applied control. The only possibility is to lead the damaged system to its optimal operating order with respect to desirable performance and their degrees of priority. The reference inputs generation, which leads the damaged system to its optimal operating order, corresponds to the nonlinear quadratic programming optimization problem. The objective is to minimize the distance between the desirable output vector y_d before failure and the desired output vector y_f, distributing the energy equally among the healthy actuators and such that (y_f, u_f) is solution of the damaged system (Σ_j) and belonging to the set (Ψ). Thus, the optimization problem is defined as

$$J_{opt}\left(y_f^*, u_f^*\right) = \min_{y_f,\ u_f} \left(\|y_d - y_f\|_{Q_{opt}^{1/2}}^2 + \|u_f\|_{R_{opt,f}^{1/2}}^2 \right), \qquad (4.54)$$

under the constraints

$$\begin{cases} f_j(x_f) + \sum_{i=1,\ i\neq j}^{m} g_i(x_f)u_{i,f} = 0, \\ y_f - h(x_f) = 0, \\ x_f - x_{min} \geq 0, x_{max} - x_f \geq 0, \\ y_f - y_{min} \geq 0, y_{max} - y_f \geq 0, \\ u_{i,f} - u_{i,min} \geq 0, u_{i,max} - u_{i,f} \geq 0, \qquad i = 1, ..., m,\ i \neq j, \end{cases} \qquad (4.55)$$

where $R_{opt,f} = \begin{bmatrix} I_{j-1} & 0 & 0 \\ 0 & 0 & I_{m-j} \end{bmatrix}$ and $Q_{opt} = \begin{bmatrix} I_{j-1} & 0 \\ 0 & 0 \\ 0 & I_{m-j} \end{bmatrix}$.

Matrices $Q_{opt} \in \Re^{m \times m}$ and $R_{opt,f} \in \Re^{m \times m}$ correspond, respectively, to priority degree of outputs and to solicitation degree of actuators.

The system can continue to operate with degraded performance as long as it remains within acceptable limits, defined by safety constraints and the threshold of the admissible lowest quality T:

$$T = \|E_{max}\|^2_{Q^{1/2}_{opt}} + \|u_{max}\|^2_{R^{1/2}_{opt,f}}, \tag{4.56}$$

where $E_{max} = |y_d - y_f|_{max}$ represents the accepted maximal output error and u_{max} the maximal energy provided by each actuator.

If $J_{opt} < T$, the reference inputs of the new control law are modified to lead the damaged system to operating order $\left(y_f^*, u_f^*\right)$ rather than stopping it.

The proposed method is based on the computation of new reference inputs in order to drive the system in an optimal operating order (optimal trim point for linear systems) with respect to desirable performances and to their degrees of priority. For a plant without actuator redundancies such as the three-tank system, this method is very important because it is impossible to maintain the system at some acceptable level of performance in the presence of major actuator failures. The next paragraph is dedicated to highlight the effectiveness of this approach to the nonlinear model of the three-tank system.

Results and Comments

When the pump 1 is blocked at its current value in steady-state at $t = 1280\ s$, this fault becomes legible and critical only when the reference input associated with ℓ_2 changes at $t = 1300\ s$ as illustrated in Fig. 4.41. Indeed, the modification of flow rate q_2 with a jammed flow rate q_1 involves the uncontrolled increase of level ℓ_1. The security system avoids the overflowing of tank 1. The system then oscillates because of the successive stops of pump 1. The tracking error on level ℓ_2 being different from zero, the flow q_2 becomes maximal (Fig. 4.42). The damaged system will generally be shut down.

For the three-tank system, the constraints are defined as follows:

$$\begin{cases} \begin{bmatrix} 0 & 0 & 0 \end{bmatrix}^T \leq x(t) \leq \begin{bmatrix} 0.635 & 0.635 & 0.635 \end{bmatrix}^T \\ \begin{bmatrix} 0 & 0 \end{bmatrix}^T \leq y(t) \leq \begin{bmatrix} 0.635 & 0.635 \end{bmatrix}^T \\ \begin{bmatrix} 0 & 0 \end{bmatrix}^T \leq u(t) \leq \begin{bmatrix} 10^{-4} & 10^{-4} \end{bmatrix}^T \end{cases} \tag{4.57}$$

In this case, the damaged system does not shut down. At $t = 2700\ s$, the reference input associated to ℓ_2 becomes equal to $0.25\ m$. When the reference value of ℓ_2 changes, ℓ_2 can be correctly tracked but without worrying about

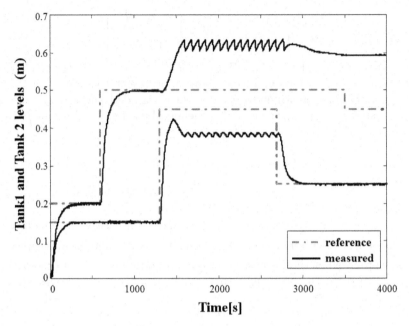

Fig. 4.41. Dynamic behavior of levels with freezing pump 1

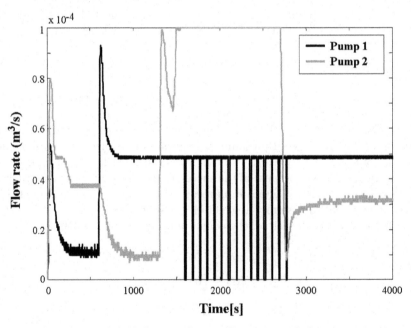

Fig. 4.42. Dynamic behavior of input flow rates with freezing pump1

level ℓ_1. This is because the nominal control is decoupled. As illustrated in Fig. 4.43, the FTC strategy avoids this critical situation and minimizes the fall of quality. Let us define the quality of product by the following matrices:

$$Q_{opt} = \begin{bmatrix} 50 & 0 \\ 0 & 25 \end{bmatrix}; \quad R_{opt} = I_2 \times 10^4; \quad E_{max} = \begin{bmatrix} 0.15 \\ 0.15 \end{bmatrix}. \quad (4.58)$$

The detection module diagnoses the pump 1 blocking at 1305 s. The threshold of the admissible lowest quality T given by (4.56), in the case of pump 1 blocking, is equal to 1.688. The new control law allows tracking of ℓ_2, and the new reference input leading the damaged system to its optimal operating order is then determined. Table 4.2 gives the operating orders reached by the nominal control and the FTC, in the presence of the major failure. It can be seen in Table 4.2 that the FTC avoids the system stop when pump 1 is blocked. The performances are degraded but remain better than the admissible lowest quality $J_{opt} < T$. Regardless of the desirable operating order, the performances are always better than without FTC.

Table 4.2. Operating orders in case of pump 1 blocked at $t = 1280$ s

y_d	Nominal control with fault		Fault-tolerant control with fault	
$[0.5 \quad 0.45]^T$	Critical		$[0.568 \quad 0.233]^T$	$J_{opt} = 1.408$
$[0.5 \quad 0.25]^T$	$[0.594 \quad 0.25]^T$	$J_{opt} = 0.442$	$[0.512 \quad 0.166]^T$	$J_{opt} = 0.184$
$[0.45 \quad 0.25]^T$	$[0.594 \quad 0.25]^T$	$J_{opt} = 1.04$	$[0.49 \quad 0.133]^T$	$J_{opt} = 0.422$

4.5 Conclusion

The three-tank system is used to illustrate the abilities of the FTC system to compensate for both sensor and actuator faults. A bank of unknown input observers has been designed in order to detect, isolate, and estimate faults, and principally to distinguish between sensor and actuator faults.

Indeed, since the compensation for an actuator fault cannot be achieved in the same way as for a sensor fault, it is of great importance to distinguish between these faults, which is not usually easy in closed-loop systems. Various experiments have been conducted in the presence of sensor and actuator faults. The fault accommodation method based on the FDI results shows that the compensated outputs behavior are closer to the nominal outputs than the faulty outputs without compensation.

The main difficulty in applying this FDI and accommodation method in more complex industrial system is the establishment of an analytical model. However in many cases, the breaking down of the whole system into subsystems makes possible the modeling and the application of such a method. It

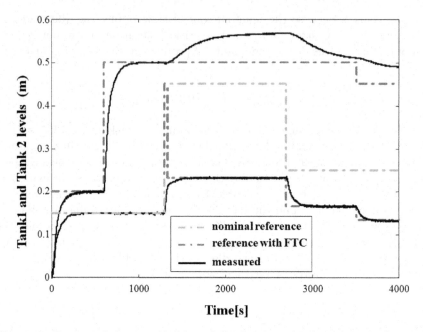

Fig. 4.43. Dynamic behavior of levels with FTC system in a freezing case of pump 1

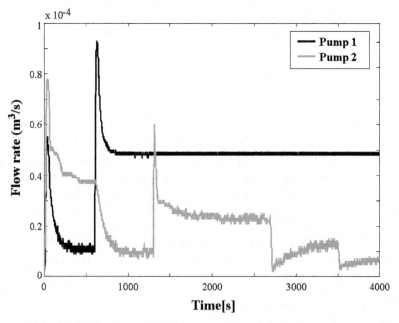

Fig. 4.44. Dynamic behavior of input flow rates with FTC system of freezing pump 1

can be noted that for each subsystem the model obtained is assumed to be linear around an operating point and is suitable in the associated operating zone.

Moreover, the nonlinear model of the system is considered with a nominal control law based on an exact linearization input-output by state feedback. Because the nonlinear features of the system are considered, the fault magnitude is estimated regardless of the operating zone. This method is suitable for actuator faults such as biases or a loss in the effectiveness of an actuator or a system component. The developed method emphasizes the importance of the FTC applied to the three-tank system. A decoupling approach and a bank of nonlinear observers has been designed in order to detect, isolate, and estimate sensor faults. Then the fault is compensated using the fault free estimation of the measured outputs. This method is suitable for sensor faults such as biases or drift of sensor gain. it should be noted that one way to consider a wide operating zone could consist in the use of multiple models techniques rather than considering an exact nonlinear model [106].

A major actuator failure (blocking or complete loss) in a process without redundancy can be considered as a critical failure. Its application to a nonlinear process, the three-tank system, emphasizes the importance and usefulness of such fault-tolerance. The objective of this FTC is different from that usually met. Indeed, in the presence of critical faults on such a process, it is impossible to maintain the damaged system at some acceptable level of performance, regardless of the applied control strategy. The objective is to operate safely and to minimize the loss of productivity. It is realized from a nonlinear quadratic programming optimization. Recent works have considered compensating a complete loss of an actuator. The occurrence of this kind of critical failure requires either a hardware redundancy or an on-line modification of the nominal objectives in order to avoid catastrophic consequences until shutting down the system safely under explicit input saturation constraints [120].

In this chapter, the robustness against modeling errors has not been considered. However, the FDI and FTC concepts are always valid in spite of error models which will only modify, for instance, the sensitivity of an unknown input observer against faults, or the error state estimator convergence. Some recent papers have considered this topic for the three-tank system. For instance, robust FDI filter design problem for LTI uncertain systems under feedback control has been investigated in [63] where two-design methods involving norm-based fault detection filters are applied to the three-tank system, and compared to each other. To deal with imprecisions and uncertainties of models, these uncertainties are generally represented using interval models. In [108] an approach is proposed to generate envelopes based on interval techniques of the modal interval analysis. In the framework of off-line model-based FDI for multi-variable uncertain systems, a method is proposed in [64] using the generalized structured singular value and based on frequency-domain model invalidation tools. Other studies considered robust FDI of nonlinear systems, as for example in [87] where new adaptive law and SMOs with boundary layer control are introduced into Polycarpou's on-line approximator to offer a fast and robust FDI strategy for a class of nonlinear systems.

As one of several popular experimental systems in control laboratories, the three-tank process is a perfect experimental process setup for investigating linear and nonlinear multi-variable feedback control as well as FDI and FTC system design. The method developed in this chapter allows increasing of the application field of FTC systems.

Sensor Fault-tolerant Control Method for Active Suspension System

5.1 Introduction

The main objective of vehicle suspension is to reduce the effect of the vibrations generated by road irregularities on the human body. The suspension system is classified as a passive, semi-active, or active suspension, according to its ability to add or extract energy. In active suspension, the force actuator is able to both add and dissipate energy from the system. This will enable the suspension to control the attitude of the vehicle, reduce the effects of braking and vehicle roll during cornering maneuvers, in addition to its capability to increase ride comfort and vehicle road handling.

The active suspension control problem has been widely studied in the literature: a state and output-feedback scheduled [84]; a modular adaptive [21]; a fuzzy logic [27,118]; an adaptive fuzzy [26]; a stochastic optimal [94]; a mixed H_2/H_∞ [51]; a proportional-integral sliding mode [109]; a combined filtered feedback controller [67]; an H_∞ [32, 33, 85, 129]; a neural network [57]; an LQ regulator [110]; a mixed LQ regulator/backstepping technique [89]; a sliding mode [134–136]; an adaptive sliding controller [23] and many other controllers were designed and applied to quarter, half, and full vehicle active suspension systems.

This chapter aims to design and integrate control, diagnosis and fault-tolerance for a nonlinear full vehicle active suspension. The system consists of the chassis and the four suspension systems. In addition, the dynamics of the four force actuators are taken into consideration. This is because force actuators are crucial elements of the system, as they have their own dynamics and they are subjected to faults. The resulting system is a nonlinear large-scale complex system of 22 states. To facilitate its study, it is broken down physically into five interconnected subsystems. Each subsystem has its own sensors and actuators.

The control law is designed using sliding mode techniques: a local control module is designed for each subsystem whereas a global control module at a higher level monitors and supervises these local modules and ensures an intelligent synchronization and coordination between the subsystems. In parallel with this structure, a fault diagnosis structure is built. A local diagnosis module detects and isolates sensor faults in each subsystem and a global module monitors the local ones.

The chapter starts by presenting the system and its model. The control law is then detailed in Sect. 5.3. In Sect. 5.4, the instrumentation problem is discussed. The design of the diagnosis and the fault-tolerant modules is addressed in Sect. 5.5. Finally, to illustrate the control, the diagnosis and the fault-tolerance, simulation results are shown and commented on in Sect. 5.6.

5.2 Full Vehicle Active Suspension System

Active suspension systems are nonlinear as is the nature of many systems generally. The linear model can replace the nonlinear one around the operating conditions. Out of this interval, the linear model is not valid and a linear representation of the system dynamics is not sufficient. Thus, a nonlinear model of the full vehicle active suspension system is considered.

5.2.1 System Description

The full vehicle model consists of the chassis (sprung mass) connected by the suspension systems to four wheels (unsprung masses). This system is illustrated in Fig. 5.1. Each suspension is modeled as a linear viscous damper, a linear spring, and a force actuator. Each wheel is modeled as a linear spring. The chassis is free to heave, pitch, and roll. The wheels are free to bounce vertically with respect to the chassis.

The dynamics of the electro-hydraulic servo valves given by Gaspar *et al.* [51] are taken into consideration in this study (Fig. 5.2). The actuators generate the actuation forces. Thus, taking their dynamics into consideration helps in better understanding the system behavior.

Each actuation module consists of a cylinder (actuator) and a four way servo valve. The function principle of the actuation module is to generate an input force u_ϑ on the spool which is able to move forward or backward. The variation of the spool position z_{v_ϑ} results in the variation of the amount of fluid entering (or leaving) the cylinder chambers, thus creating a pressure drop across the piston and generating the actuator force f_ϑ. $\vartheta = \{fr, fl, rr, rl\}$ stands respectively for front right, front left, rear right, and rear left.

5.2.2 System Modeling

In this section, the dynamics of the sprung mass, the unsprung masses, and the actuators are illustrated. It should be noted that this chapter deals with continuous-time systems but the time will be omitted for simplification.

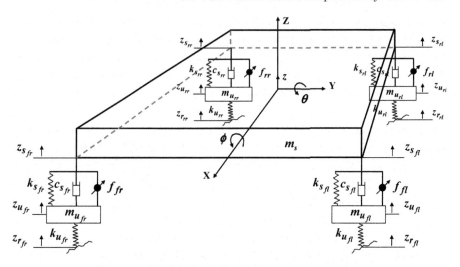

Fig. 5.1. Model of a full vehicle active suspension

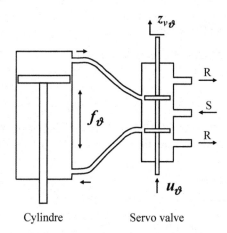

Fig. 5.2. Representation of an actuator ($\vartheta \in \{fr, fl, rr, rl\}$)

Sprung Mass Dynamics

Using Newton's law, the sprung mass (chassis) dynamics are [56]

$$
\begin{aligned}
\ddot{z} = &\{k_{s_{fr}} z_{u_{fr}} + k_{s_{fl}} z_{u_{fl}} + k_{s_{rr}} z_{u_{rr}} + k_{s_{rl}} z_{u_{rl}} - (k_{s_{fr}} + k_{s_{fl}} + k_{s_{rr}} + k_{s_{rl}}) z - \\
&[a(k_{s_{fr}} + k_{s_{fl}}) - b(k_{s_{rr}} + k_{s_{rl}})] \sin\theta - [d(k_{s_{fl}} + k_{s_{rl}}) - c(k_{s_{fr}} + k_{s_{rr}})] \sin\phi + \\
&c_{s_{fr}} \dot{z}_{u_{fr}} + c_{s_{fl}} \dot{z}_{u_{fl}} + c_{s_{rr}} \dot{z}_{u_{rr}} + c_{s_{rl}} \dot{z}_{u_{rl}} - (c_{s_{fr}} + c_{s_{fl}} + c_{s_{rr}} + c_{s_{rl}}) \dot{z} - [a(c_{s_{fr}} + \\
&c_{s_{fl}}) - b(c_{s_{rr}} + c_{s_{rl}})] \cos\theta \dot{\theta} - [d(c_{s_{fl}} + c_{s_{rl}}) - c(c_{s_{fr}} + c_{s_{rr}})] \cos\phi \dot{\phi} + f_{fr} + f_{fl} + \\
&f_{rr} + f_{rl}\}/M,
\end{aligned}
$$

$$\ddot{\theta} = \cos\theta\{ak_{s_{fr}}z_{u_{fr}} + ak_{s_{fl}}z_{u_{fl}} - bk_{s_{rr}}z_{u_{rr}} - bk_{s_{rl}}z_{u_{rl}} - [a(k_{s_{fr}} + k_{s_{fl}}) - b(k_{s_{rr}} + k_{s_{rl}})]z - [a^2(k_{s_{fr}} + k_{s_{fl}}) + b^2(k_{s_{rr}} + k_{s_{rl}})]\sin\theta - [d(ak_{s_{fl}} - bk_{s_{rl}}) - c(ak_{s_{fr}} - bk_{s_{rr}})]\sin\phi + ac_{s_{fr}}\dot{z}_{u_{fr}} + ac_{s_{fl}}\dot{z}_{u_{fl}} - bc_{s_{rr}}\dot{z}_{u_{rr}} - bc_{s_{rl}}\dot{z}_{u_{rl}} - [a(c_{s_{fr}} + c_{s_{fl}}) - b(c_{s_{rr}} + c_{s_{rl}})]\dot{z} - [a^2(c_{s_{fr}} + c_{s_{fl}}) + b^2(c_{s_{rr}} + c_{s_{rl}})]\cos\theta\dot{\theta} - [d(ac_{s_{fl}} - bc_{s_{rl}}) - c(ac_{s_{fr}} - bc_{s_{rr}})]\cos\phi\dot{\phi} + a(f_{fr} + f_{fl}) - b(f_{rr} + f_{rl})\}/I_{yy},$$

$$\ddot{\phi} = \cos\phi\{-ck_{s_{fr}}z_{u_{fr}} + dk_{s_{fl}}z_{u_{fl}} - ck_{s_{rr}}z_{u_{rr}} + dk_{s_{rl}}z_{u_{rl}} - [d(k_{s_{fl}} + k_{s_{rl}}) - c(k_{s_{fr}} + k_{s_{rr}})]z - [d(ak_{s_{fl}} - bk_{s_{rl}}) - c(ak_{s_{fr}} - bk_{s_{rr}})]\sin\theta - [d^2(k_{s_{fl}} + k_{s_{rl}}) + c^2(k_{s_{fr}} + k_{s_{rr}})]\sin\phi - cac_{s_{fr}}\dot{z}_{u_{fr}} + dc_{s_{fl}}\dot{z}_{u_{fl}} - cc_{s_{rr}}\dot{z}_{u_{rr}} + dc_{s_{rl}}\dot{z}_{u_{rl}} - [d(c_{s_{fl}} + c_{s_{rl}}) - c(c_{s_{fr}} + c_{s_{rr}})]\dot{z} - [d(ac_{s_{fl}} - bc_{s_{rl}}) - c(ac_{s_{fr}} - bc_{s_{rr}})]\cos\theta\dot{\theta} - [d^2(c_{s_{fl}} + c_{s_{rl}}) + c^2(c_{s_{fr}} + c_{s_{rr}})]\cos\phi\dot{\phi} - c(f_{fr} + f_{rr}) + d(f_{fl} + f_{rl})\}/I_{xx},$$

where z is the heave position and θ and ϕ are respectively the pitch and roll angles of the sprung mass. The other variables are illustrated in Fig. 5.1 and the different constants are defined in the sequel.

Unsprung Masses Dynamics

The unsprung masses (tires) dynamics are [56]

$$\ddot{z}_{u_\vartheta} = [k_{s_\vartheta}(z_{s_\vartheta} - z_{u_\vartheta}) + c_{s_\vartheta}(\dot{z}_{s_\vartheta} - \dot{z}_{u_\vartheta}) + k_{u_\vartheta}(z_{r_\vartheta} - z_{u_\vartheta}) - f_\vartheta]/m_{u_\vartheta}, \quad (5.1)$$

where z_{u_ϑ} ($\vartheta \in \{fr, \, fl, \, rr, \, rl\}$) is the vertical displacement of the unsprung mass. The other variables are illustrated in Fig. 5.1 and the constants are defined in the sequel.

Actuators Dynamics

The cylinder dynamics are given by [51]

$$\frac{V_t}{4\beta_e}\dot{A}_\vartheta = Q - C_{tp}A_\vartheta - S(\dot{z}_{s_\vartheta} - \dot{z}_{u_\vartheta}), \quad (5.2)$$

where A_ϑ is the actuator load pressure, V_t is the total actuator volume, β_e is the effective bulk modulus, C_{tp} is the coefficient of total leakage due to pressure, and S is the actuator ram area. $\dot{z}_{s_\vartheta}$ and $\dot{z}_{u_\vartheta}$ are, respectively, the vertical velocities of the sprung mass and the unsprung mass (see Fig. 5.1). Q, the load flow, is given by

$$Q = sign(p_s - sign(z_{v_\vartheta})A_\vartheta)C_d\omega z_{v_\vartheta}\sqrt{\frac{|p_s - sign(z_{v_\vartheta})A_\vartheta|}{\rho}}, \quad (5.3)$$

where p_s is the supply pressure, z_{v_ϑ} is the spool valve position (see Fig. 5.2), C_d is the discharge coefficient, ω is the spool valve area gradient, and ρ is the hydraulic fluid density. By replacing the load flow Q given by (5.3) in (5.2), the following equation is obtained:

$$\frac{V_t}{4\beta_e}\dot{A}_{\vartheta\vartheta} = -C_{tp}A_\vartheta - S(\dot{z}_{s_\vartheta} - \dot{z}_{u_\vartheta}) + S_g C_d \omega z_{v_\vartheta}\sqrt{\frac{|p_s - sign(z_{v_\vartheta})A_\vartheta|}{\rho}}, \quad (5.4)$$

or

$$\dot{A}_\vartheta = -\frac{4\beta_e C_{tp}}{V_t}A_\vartheta - \frac{4\beta_e}{V_t}S(\dot{z}_{s_\vartheta} - \dot{z}_{u_\vartheta}) + \frac{4\beta_e}{V_t}S_g C_d \omega z_{v_\vartheta}\sqrt{\frac{|p_s - sign(z_{v_\vartheta})A_\vartheta|}{\rho}}, \quad (5.5)$$

where $S_g = sign(p_s - sign(z_{v_\vartheta})A_\vartheta)$.

By defining α, β, and γ as $\alpha = \frac{4\beta_e}{V_t}$, $\beta = \alpha C_{tp}$ and $\gamma = \alpha C_d \omega\sqrt{\frac{1}{\rho}}$, (5.5) becomes

$$\dot{A}_\vartheta = -\beta A_\vartheta - \alpha S(\dot{z}_{s_\vartheta} - \dot{z}_{u_\vartheta}) + sign(p_s - sign(z_{v_\vartheta})A_\vartheta)\gamma\sqrt{|p_s - sign(z_{v_\vartheta})A_\vartheta|}\, z_{v_\vartheta}. \quad (5.6)$$

In [2, 3], the cylinder dynamics are given by

$$\dot{A}_\vartheta = -\beta A_\vartheta - \alpha S(\dot{z}_{s_\vartheta} - \dot{z}_{u_\vartheta}) + \gamma\sqrt{p_s - sign(z_{v_\vartheta})A_\vartheta}\, z_{v_\vartheta}. \quad (5.7)$$

In this chapter, the dynamics given by (5.6) are used. The spool valve position z_{v_ϑ} is controlled by the control input u_ϑ (see Fig. 5.2). The spool valve dynamics are modeled as first order system [51]:

$$\dot{z}_{v_\vartheta} = \frac{1}{\tau}(-z_{v_\vartheta} + u_\vartheta), \quad (5.8)$$

where τ is the time constant. The control input u_ϑ is given by

$$u_\vartheta = k i_\vartheta, \quad (5.9)$$

where i_ϑ is the servo valve current and k is the valve gain.

In conclusion, the actuator dynamics are given by

$$\dot{A}_\vartheta = -\beta A_\vartheta - \alpha S(\dot{z}_{s_\vartheta} - \dot{z}_{u_\vartheta}) + sign(p_s - sign(z_{v_\vartheta})A_\vartheta)\gamma\sqrt{|p_s - sign(z_{v_\vartheta})A_\vartheta|}\, z_{v_\vartheta} \quad (5.10)$$

and

$$\dot{z}_{v_\vartheta} = \frac{1}{\tau}(-z_{v_\vartheta} + u_\vartheta). \quad (5.11)$$

The force generated by the actuator is given by $f_\vartheta = SA_\vartheta$ where S is the actuator ram area and $\vartheta = \{fr,\ fl,\ rr,\ rl\}$.

5.2.3 System's Model

The concatenation of the vehicle active suspension and the actuator models results in an input affine nonlinear model of order 22, which can be written as a state-space representation

$$\begin{cases} \dot{x}(t) = f(x(t)) + Bu(t) + Fd(t) \\ y(t) = Cx(t) \end{cases}.$$ (5.12)

$x \in \Re^{22}$ is the state vector with:

$x_{4.k+1} = z_{u_\vartheta}$:	unsprung mass displacement
$x_{4.k+2} = \dot{z}_{u_\vartheta}$:	unsprung mass velocity
$x_{4.k+3} = A_\vartheta$:	actuator load pressure
$x_{4.k+4} = z_{v_\vartheta}$:	spool valve position
$x_{17} = z$:	heave position of the sprung mass
$x_{18} = \dot{z}$:	heave velocity of the sprung mass
$x_{19} = \theta$:	pitch angle of the sprung mass
$x_{20} = \dot{\theta}$:	pitch angular velocity of the sprung mass
$x_{21} = \phi$:	roll angle of the sprung mass
$x_{22} = \dot{\phi}$:	roll angular velocity of the sprung mass

where $(k, \vartheta) = \{(0, fr), (1, fl), (2, rr), (3, rl)\}$. The dot "." represents multiplication. $u(t) = [u_{fr}(t) \quad u_{fl}(t) \quad u_{rr}(t) \quad u_{rl}(t)]^T \in \Re^4$ is the vector of the control inputs to the four servo valves. $d(t) = [z_{r_{fr}}(t) \quad z_{r_{fl}}(t) \quad z_{r_{rr}}(t) \quad z_{r_{rl}}(t)]^T \in \Re^4$ is the vector of the external disturbances induced to the active suspension due to the road irregularities.

The output vector $y(t)$ is discussed in Sect. 5.4 where the sensors needed for the system control are presented and their existence is investigated. The function $f(x)$, matrices B and F, and the numerical values of the different parameters are given in Sect. 5.2.5.

5.2.4 System Breakdown

Complex systems are quantitatively characterized by a large dimension of the mathematical model, a large number of input and output variables. Qualitatively, the complexity consists of the system nonlinearity, external disturbances, structural and parametric uncertainty, and sophisticated and multiple objectives and performance criteria. When controlling a complex system, the desired control law cannot be derived from the model of the whole system without some kind of transformation or simplification. The most common approaches used to solve the problem of complexity include various techniques of simplification and breakdown. Breakdown techniques can be classified as follows [42]:

- Task-oriented (functional) associated with splitting the general problem into several sub-problems of various hierarchical levels

- Object-oriented (physical) implying separation of simpler subsystems to be considered individually
- Time breakdown or separating distinct stages of system functioning

Vehicles are complex systems which consist of various mechanical, electronic, and electrical subsystems. The full vehicle active suspension model considered in this study is a deterministic and simplified representation of the real complex system. Nevertheless, it is a large dimension nonlinear system with a relatively large number of input and output variables. It has sophisticated and multiple objectives and it is subjected to external disturbances. Thus, it is considered as a complex system. To facilitate the study, the system is physically broken down into five interconnected subsystems to be considered separately as shown in Fig. 5.3.

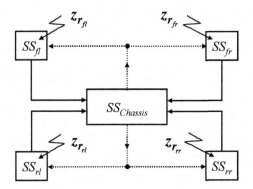

Fig. 5.3. Breakdown of the system into five interconnected subsystems

The arrows in the figure represent the interconnections between subsystems. This breakdown respects the system's physical structure. The five subsystems are the front right SS_{fr}, the front left SS_{fl}, the rear right SS_{rr}, the rear left SS_{rl}, and the chassis $SS_{chassis}$. The breakdown facilitates the design of the controller and the FDI modules.

For each subsystem SS_{ϑ}, a local controller C_{ϑ} and an FDI module $Diag_{\vartheta}$ ($\vartheta \in \{fr, fl, rr, rl, chassis\}$) are designed. The subsystems with their control and diagnosis modules are illustrated in Fig. 5.4.

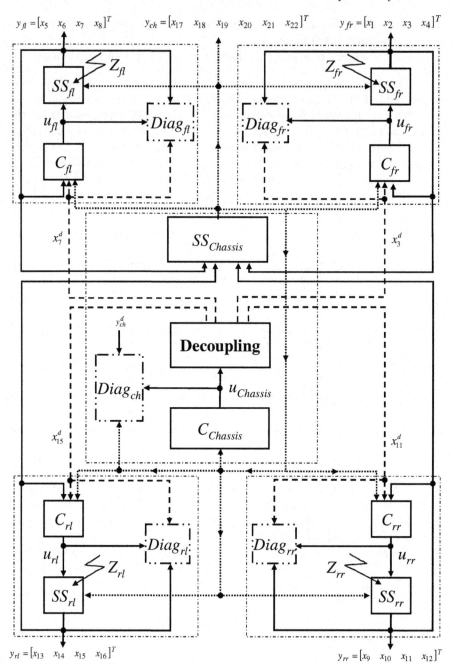

Fig. 5.4. Local control and FDI modules for subsystems

Figure 5.5 illustrates the functional breakdown of control and diagnosis structures. The local control modules C_ϑ are monitored by a *coordination and synchronization* module at a higher level. At the top of this control structure, a *supervision* module supervises all the sub-modules. In parallel with this hierarchical structure of control, a fault diagnosis scheme is designed: a *functional diagnosis* module at level 1 coordinates the local FDI modules $Diag_\vartheta$. A *global diagnosis* module at the top of this hierarchy supervises all these sub-modules. The coordination between the two structures is ensured by the *decision taking and resources management* module.

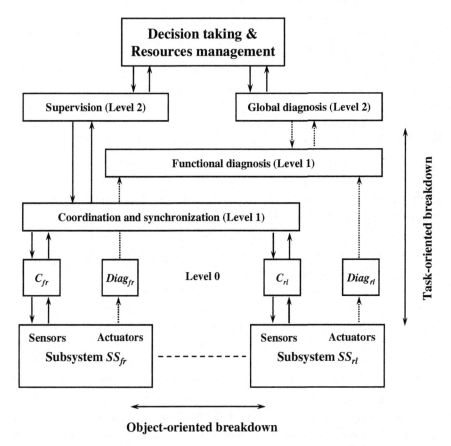

Fig. 5.5. Object and task-oriented breakdown of the system

5.2.5 Subsystems Models and Vehicle Parameters

The dynamics of each subsystem are the following.

Front Right Subsystem SS_{fr}

$$f_1(x) = x_2,$$

$$f_2(x) = [-(k_{s_{fr}} + k_{u_{fr}})x_1 + k_{s_{fr}}x_{17} + ak_{s_{fr}}\sin(x_{19}) - ck_{s_{fr}}\sin(x_{21}) - c_{s_{fr}}x_2 + c_{s_{fr}}x_{18} + ac_{s_{fr}}\cos(x_{19})x_{20} - cc_{s_{fr}}\cos(x_{21})x_{22} - Sx_3]/m_{u_{fr}},$$

$$f_3(x) = -\beta x_3 - \alpha S[x_{18} + ax_{20}\cos(x_{19}) - cx_{22}\cos(x_{21}) - x_2] + sign(p_s - sign(x_4)x_3)\gamma\sqrt{|p_s - sign(x_4)x_3|}x_4,$$

$$f_4(x) = -x_4/\tau.$$

Front Left Subsystem SS_{fl}

$$f_5(x) = x_6,$$

$$f_6(x) = [-(k_{s_{fl}} + k_{u_{fl}})x_5 + k_{s_{fl}}x_{17} + ak_{s_{fl}}\sin(x_{19}) + dk_{s_{fl}}\sin(x_{21}) - c_{s_{fl}}x_6 + c_{s_{fl}}x_{18} + ac_{s_{fl}}\cos(x_{19})x_{20} + dc_{s_{fl}}\cos(x_{21})x_{22} - Sx_7]/m_{u_{fl}},$$

$$f_7(x) = -\beta x_7 - \alpha S[x_{18} + ax_{20}\cos(x_{19}) + dx_{22}\cos(x_{21}) - x_6] + sign(p_s - sign(x_8)x_7)\gamma\sqrt{|p_s - sign(x_8)x_7|}x_8,$$

$$f_8(x) = -x_8/\tau.$$

Rear Right Subsystem SS_{rr}

$$f_9(x) = x_{10},$$

$$f_{10}(x) = [-(k_{s_{rr}} + k_{u_{rr}})x_9 + k_{s_{rr}}x_{17} - bk_{s_{rr}}\sin(x_{19}) - ck_{s_{rr}}\sin(x_{21}) - c_{s_{rr}}x_{10} + c_{s_{rr}}x_{18} - bc_{s_{rr}}\cos(x_{19})x_{20} - cc_{s_{rr}}\cos(x_{21})x_{22} - Sx_{11}]/m_{u_{rr}},$$

$$f_{11}(x) = -\beta x_{11} - \alpha S[x_{18} - bx_{20}\cos(x_{19}) - cx_{22}\cos(x_{21}) - x_{10}] + sign(p_s - sign(x_{12})x_{11})\gamma\sqrt{|p_s - sign(x_{12})x_{11}|}x_{12},$$

$$f_{12}(x) = -x_{12}/\tau.$$

Rear Left Subsystem SS_{rl}

$$f_{13}(x) = x_{14},$$

$$f_{14}(x) = [-(k_{s_{rl}} + k_{u_{rl}})x_{13} + k_{s_{rl}}x_{17} - bk_{s_{rl}}\sin(x_{19}) + dk_{s_{rl}}\sin(x_{21}) - c_{s_{rl}}x_{14} + c_{s_{rl}}x_{18} - bc_{s_{rl}}\cos(x_{19})x_{20} + dc_{s_{rl}}\cos(x_{21})x_{22} - Sx_{15}]/m_{u_{rl}},$$

$$f_{15}(x) = -\beta x_{15} - \alpha S[x_{18} - bx_{20}\cos(x_{19}) + dx_{22}\cos(x_{21}) - x_{14}] + sign(p_s - sign(x_{16})x_{15})\gamma\sqrt{|p_s - sign(x_{16})x_{15}|x_{16}},$$

$$f_{16}(x) = -x_{16}/\tau.$$

Chassis Subsystem $SS_{chassis}$

$$f_{17}(x) = x_{18},$$

$$f_{18}(x) = \{k_{s_{fr}}x_1 + k_{s_{fl}}x_5 + k_{s_{rr}}x_9 + k_{s_{rl}}x_{13} - (k_{s_{fr}} + k_{s_{fl}} + k_{s_{rr}} + k_{s_{rl}})x_{17} - [a(k_{s_{fr}} + k_{s_{fl}}) - b(k_{s_{rr}} + k_{s_{rl}})]\sin(x_{19}) - [d(k_{s_{fl}} + k_{s_{rl}}) - c(k_{s_{fr}} + k_{s_{rr}})]\sin(x_{21}) + c_{s_{fr}}x_2 + c_{s_{fl}}x_6 + c_{s_{rr}}x_{10} + c_{s_{rl}}x_{14} - (c_{s_{fr}} + c_{s_{fl}} + c_{s_{rr}} + c_{s_{rl}})x_{18} - [a(c_{s_{fr}} + c_{s_{fl}}) - b(c_{s_{rr}} + c_{s_{rl}})]\cos(x_{19})x_{20} - [d(c_{s_{fl}} + c_{s_{rl}}) - c(c_{s_{fr}} + c_{s_{rr}})]\cos(x_{21})x_{22} + S(x_3 + x_7 + x_{11} + x_{15})\}/M,$$

$$f_{19}(x) = x_{20},$$

$$f_{20}(x) = \cos(x_{19})\{ak_{s_{fr}}x_1 + ak_{s_{fl}}x_5 - bk_{s_{rr}}x_9 - bk_{s_{rl}}x_{13} - [a(k_{s_{fr}} + k_{s_{fl}}) - b(k_{s_{rr}} + k_{s_{rl}})]x_{17} - [a^2(k_{s_{fr}} + k_{s_{fl}}) + b^2(k_{s_{rr}} + k_{s_{rl}})]\sin(x_{19}) - [d(ak_{s_{fl}} - bk_{s_{rl}}) - c(ak_{s_{fr}} - bk_{s_{rr}})]\sin(x_{21}) + ac_{s_{fr}}x_2 + ac_{s_{fl}}x_6 - bc_{s_{rr}}x_{10} - bc_{s_{rl}}x_{14} - [a(c_{s_{fr}} + c_{s_{fl}}) - b(c_{s_{rr}} + c_{s_{rl}})]x_{18} - [a^2(c_{s_{fr}} + c_{s_{fl}}) + b^2(c_{s_{rr}} + c_{s_{rl}})]\cos(x_{19})x_{20} - [d(ac_{s_{fl}} - bc_{s_{rl}}) - c(ac_{s_{fr}} - bc_{s_{rr}})]\cos(x_{21})x_{22} + S[a(x_3 + x_7) - b(x_{11} + x_{15})]\}/I_{yy},$$

$$f_{21}(x) = x_{22},$$

$$f_{22}(x) = \cos(x_{21})\{-ck_{s_{fr}}x_1 + dk_{s_{fl}}x_5 - ck_{s_{rr}}x_9 + dk_{s_{rl}}x_{13} - [d(k_{s_{fl}} + k_{s_{rl}}) - c(k_{s_{fr}} + k_{s_{rr}})]x_{17} - [d(ak_{s_{fl}} - bk_{s_{rl}}) - c(ak_{s_{fr}} - bk_{s_{rr}})]\sin(x_{19}) - [d^2(k_{s_{fl}} + k_{s_{rl}}) + c^2(k_{s_{fr}} + k_{s_{rr}})]\sin(x_{21}) - cac_{s_{fr}}x_2 + dc_{s_{fl}}x_6 - cc_{s_{rr}}x_{10} + dc_{s_{rl}}x_{14} - [d(c_{s_{fl}} + c_{s_{rl}}) - c(c_{s_{fr}} + c_{s_{rr}})]x_{18} - [d(ac_{s_{fl}} - bc_{s_{rl}}) - c(ac_{s_{fr}} - bc_{s_{rr}})]\cos(x_{19})x_{20} - [d^2(c_{s_{fl}} + c_{s_{rl}}) + c^2(c_{s_{fr}} + c_{s_{rr}})]\cos(x_{21})x_{22} + S[-c(x_3 + x_{11}) + d(x_7 + x_{15})]\}/I_{xx}.$$

Matrices B and F are given as

$$B = \begin{bmatrix} 0_{3\times4} \\ 1/\tau & 0 & 0 & 0 \\ 0_{3\times4} \\ 0 & 1/\tau & 0 & 0 \\ 0_{3\times4} \\ 0 & 0 & 1/\tau & 0 \\ 0_{3\times4} \\ 0 & 0 & 0 & 1/\tau \\ 0_{6\times4} \end{bmatrix}, F = \begin{bmatrix} 0_{1\times4} \\ k_{u_{fr}}/m_{u_{fr}} & 0 & 0 & 0 \\ 0_{3\times4} \\ 0 & k_{u_{fl}}/m_{u_{fl}} & 0 & 0 \\ 0_{3\times4} \\ 0 & 0 & k_{u_{rr}}/m_{u_{rr}} & 0 \\ 0_{3\times4} \\ 0 & 0 & 0 & k_{u_{rl}}/m_{u_{rl}} \\ 0_{8\times4} \end{bmatrix}.$$

Expression $x_{18} + ax_{20}\cos(x_{19}) - cx_{22}\cos(x_{21}) - x_2$ in $f_3(x)$ denotes the front right suspension stroke rate $\dot{z}_{s_{fr}} - \dot{z}_{u_{fr}}$ but written as function of the system state. This is similar for $f_7(x)$, $f_{11}(x)$, and $f_{15}(x)$.

The numerical values of the different system parameters are given in Table 5.1, where G is the center of gravity of the chassis and $j = \{r,\ l\}$.

Table 5.1. System parameters

Parameter	Description	Value	Unit
M	Sprung mass	1,500	$[kg]$
m_{u_ϑ}	Unsprung masses	59	$[kg]$
$k_{s_{fj}}$	Front springs stiffness	35,000	$[N/m]$
$k_{s_{rj}}$	Rear springs stiffness	38,000	$[N/m]$
k_{u_ϑ}	Tires stiffness	190,000	$[N/m]$
$c_{s_{fj}}$	Front suspensions damping	1,000	$[N/m/s]$
$c_{s_{rj}}$	Rear suspensions damping	1,100	$[N/m/s]$
I_{xx}	Roll axis moment of inertia	460	$[kg.m^2]$
I_{yy}	Pitch axis moment of inertia	2,160	$[kg.m^2]$
a	G to the front axle	1.4	$[m]$
b	G to the rear axle	1.7	$[m]$
c	G to the vehicle right side	1.5	$[m]$
d	G to the vehicle left side	1.5	$[m]$
α	Actuator parameter	4.515×10^{13}	$[N/m^5]$
β	Actuator parameter	1	–
γ	Actuator parameter	1.545×10^9	$[N/m^{5/2}/kg^{1/2}]$
p_s	Supply pressure	10,342,500	$[Pa]$
S	Sect.al area of the piston	3.35×10^{-4}	$[m^2]$
τ	Servo valve time constant	0.003	$[s]$

5.3 Controller Design

The concept of the sliding mode control (SMC) technique presented in Sect. 2.4.2 is applied to the vehicle active suspension. The main objective of the control is to reduce the effect of the road irregularities on the passengers and to insure the system safety during vehicle maneuvers. A control strategy (Fig. 5.6) is designed for the system using SMC techniques. The control law cannot be directly derived from the complex model of the system. Thus, the controller is functionally broken down into five modules C_ϑ, one for each subsystem SS_ϑ ($\vartheta \in \{fr,\ fl,\ rr,\ rl,\ chassis\}$). The module $C_{chassis}$ supervises the chassis state and compares it to the objectives. It then determines the necessary (desired) forces u_1^d to attain the desired objectives. A *decoupling* block calculates from the desired forces u_1^d the four actuation forces f_ϑ^d that should be generated by the four actuators.

Each f_ϑ^d is treated by the control module C_ϑ ($\vartheta \in \{fr,\ fl,\ rr,\ rl\}$). This module is composed of two sub-modules C_ϑ^c and C_ϑ^s ($\vartheta \in \{fr,\ fl,\ rr,\ rl\}$). C_ϑ^c is proper to the cylinders whereas C_ϑ^s is proper to the servo valves. Each submodule C_ϑ^c treats the desired force f_ϑ^d and determines the necessary position $z_{v_\vartheta}^d$ of the spool that generates the desired force. Finally, C_ϑ^s treats the desired position $z_{v_\vartheta}^d$ and generates the control input u_ϑ that moves the spool to the desired position. The application of the control input u_ϑ on the spool varies its position z_{v_ϑ} and generates the actuation force f_ϑ. Ideally, the force f_ϑ generated by the actuator equals the desired one f_ϑ^d ($f_\vartheta^d = f_\vartheta$). The dynamics of the three control modules are detailed in the sequel.

5.3.1 First Control Module: Chassis Module $C_{Chassis}$

Consider the six equations of $SS_{chassis}$ (see Sect. 5.2.5) and define u_z, u_θ and u_ϕ as

$$\begin{cases} u_z = S(x_3 + x_7 + x_{11} + x_{15}) \\ u_\theta = S[a(x_3 + x_7) - b(x_{11} + x_{15})] \\ u_\phi = S[-c(x_3 + x_{11}) + d(x_7 + x_{15})] \end{cases}, \qquad (5.13)$$

where S, a, b, c, and d are given in Table 5.1. The six equations of $SS_{chassis}$ can be written as

$$\begin{cases} f_{17}(x) = x_{18}, \\ f_{18}(x) = \bar{f}_{18}(x) + g_{18}u_z, \\ f_{19}(x) = x_{20}, \\ f_{20}(x) = \bar{f}_{20}(x) + g_{20}u_\theta, \\ f_{21}(x) = x_{22}, \\ f_{22}(x) = \bar{f}_{22}(x) + g_{22}u_\phi \end{cases}, \qquad (5.14)$$

with

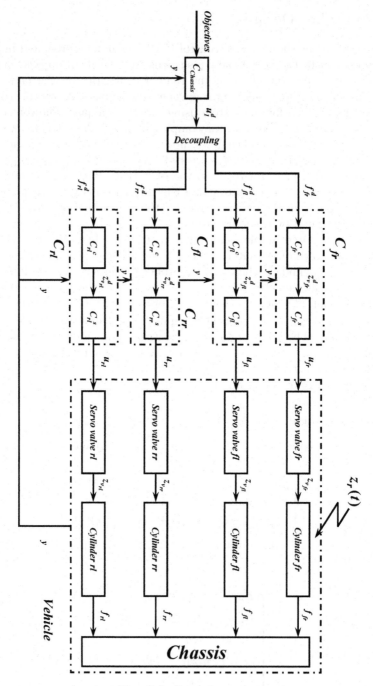

Fig. 5.6. Control strategy

$$\bar{f}_{18}(x) = \{k_{s_{fr}}x_1 + k_{s_{fl}}x_5 + k_{s_{rr}}x_9 + k_{s_{rl}}x_{13} - (k_{s_{fr}} + k_{s_{fl}} + k_{s_{rr}} + k_{s_{rl}})x_{17} -$$
$$[a(k_{s_{fr}} + k_{s_{fl}}) - b(k_{s_{rr}} + k_{s_{rl}})]\sin(x_{19}) - [d(k_{s_{fl}} + k_{s_{rl}}) - c(k_{s_{fr}} + k_{s_{rr}})]\sin(x_{21}) +$$
$$c_{s_{fr}}x_2 + c_{s_{fl}}x_6 + c_{s_{rr}}x_{10} + c_{s_{rl}}x_{14} - (c_{s_{fr}} + c_{s_{fl}} + c_{s_{rr}} + c_{s_{rl}})x_{18} - [a(c_{s_{fr}} + c_{s_{fl}}) -$$
$$b(c_{s_{rr}} + c_{s_{rl}})]\cos(x_{19})x_{20} - [d(c_{s_{fl}} + c_{s_{rl}}) - c(c_{s_{fr}} + c_{s_{rr}})]\cos(x_{21})x_{22}\}/M,$$

$$g_{18} = 1/M,$$

$$\bar{f}_{20}(x) = \cos(x_{19})\{ak_{s_{fr}}x_1 + ak_{s_{fl}}x_5 - bk_{s_{rr}}x_9 - bk_{s_{rl}}x_{13} - [a(k_{s_{fr}} + k_{s_{fl}}) -$$
$$b(k_{s_{rr}} + k_{s_{rl}})]x_{17} - [a^2(k_{s_{fr}} + k_{s_{fl}}) + b^2(k_{s_{rr}} + k_{s_{rl}})]\sin(x_{19}) - [d(ak_{s_{fl}} - bk_{s_{rl}}) -$$
$$c(ak_{s_{fr}} - bk_{s_{rr}})]\sin(x_{21}) + ac_{s_{fr}}x_2 + ac_{s_{fl}}x_6 - bc_{s_{rr}}x_{10} - bc_{s_{rl}}x_{14} - [a(c_{s_{fr}} +$$
$$c_{s_{fl}}) - b(c_{s_{rr}} + c_{s_{rl}})]x_{18} - [a^2(c_{s_{fr}} + c_{s_{fl}}) + b^2(c_{s_{rr}} + c_{s_{rl}})]\cos(x_{19})x_{20} - [d(ac_{s_{fl}} -$$
$$bc_{s_{rl}}) - c(ac_{s_{fr}} - bc_{s_{rr}})]\cos(x_{21})x_{22}\}/I_{yy},$$

$$g_{20} = \cos(x_{19})/I_{yy},$$

$$\bar{f}_{22}(x) = \cos(x_{21})\{-ck_{s_{fr}}x_1 + dk_{s_{fl}}x_5 - ck_{s_{rr}}x_9 + dk_{s_{rl}}x_{13} - [d(k_{s_{fl}} + k_{s_{rl}}) -$$
$$c(k_{s_{fr}} + k_{s_{rr}})]x_{17} - [d(ak_{s_{fl}} - bk_{s_{rl}}) - c(ak_{s_{fr}} - bk_{s_{rr}})]\sin(x_{19}) - [d^2(k_{s_{fl}} + k_{s_{rl}}) +$$
$$c^2(k_{s_{fr}} + k_{s_{rr}})]\sin(x_{21}) - cac_{s_{fr}}x_2 + dc_{s_{fl}}x_6 - cc_{s_{rr}}x_{10} + dc_{s_{rl}}x_{14} - [d(c_{s_{fl}} +$$
$$c_{s_{rl}}) - c(c_{s_{fr}} + c_{s_{rr}})]x_{18} - [d(ac_{s_{fl}} - bc_{s_{rl}}) - c(ac_{s_{fr}} - bc_{s_{rr}})]\cos(x_{19})x_{20} -$$
$$[d^2(c_{s_{fl}} + c_{s_{rl}}) + c^2(c_{s_{fr}} + c_{s_{rr}})]\cos(x_{21})x_{22}\}/I_{xx},$$

$$g_{22} = \cos(x_{21})/I_{xx}.$$

If, for the set of the first two equations $f_{17}(x)$ and $f_{18}(x)$, the desired trajectory is defined as x_{17}^d, then the error between the actual and the desired trajectory can be written as

$$e_z = z - z^d = x_{17} - x_{17}^d. \tag{5.15}$$

The time derivative of e_z is given by

$$\dot{e}_z = \dot{x}_{17} - \dot{x}_{17}^d = x_{18} - x_{18}^d. \tag{5.16}$$

The switching (sliding) surface s_z is defined as

$$s_z = \dot{e}_z + \lambda_z e_z. \tag{5.17}$$

The time derivative of s_z is given by

$$\dot{s}_z = \ddot{e}_z + \lambda_z \dot{e}_z = \bar{f}_{18}(x) + g_{18}u_z - \ddot{x}_{17}^d + \lambda_z \dot{e}_z. \tag{5.18}$$

Then the equivalent control is chosen as

$$u_{z_{eq}} = g_{18}^{-1}(\ddot{x}_{17}^d - \bar{f}_{18}(x) - \lambda_z \dot{e}_z). \tag{5.19}$$

The proportional $(k_z s_z)$ rate reaching law is imposed by selecting the second term as

$$u_z^* = g_{18}^{-1}(-k_z s_z). \tag{5.20}$$

The control input u_z then becomes

$$u_z = u_{z_{eq}} + u_z^* = g_{18}^{-1}(\ddot{x}_{17}^d - \bar{f}_{18}(x) - \lambda_z \dot{e}_z - k_z s_z). \tag{5.21}$$

By following the same steps for the set of equations $(f_{19}(x), f_{20}(x))$ and $(f_{21}(x), f_{22}(x))$, two equations are obtained:

$$u_\theta = u_{\theta_{eq}} + u_\theta^* = g_{20}^{-1}(\ddot{x}_{19}^d - \bar{f}_{20}(x) - \lambda_\theta \dot{e}_\theta - k_\theta s_\theta) \tag{5.22}$$

and

$$u_\phi = u_{\phi_{eq}} + u_\phi^* = g_{22}^{-1}(\ddot{x}_{21}^d - \bar{f}_{22}(x) - \lambda_\phi \dot{e}_\phi - k_\phi s_\phi). \tag{5.23}$$

In matrix form:

$$u_1^d = G^{-1}(\ddot{x}^d - \bar{f}(x) - \Lambda \dot{e} - ks), \tag{5.24}$$

where

$$u_1^d = \begin{bmatrix} u_z^d \\ u_\theta^d \\ u_\phi^d \end{bmatrix}, G^{-1} = \begin{bmatrix} g_{18}^{-1} & 0 & 0 \\ 0 & g_{20}^{-1} & 0 \\ 0 & 0 & g_{22}^{-1} \end{bmatrix}, \ddot{x}^d = \begin{bmatrix} \ddot{x}_{17}^d \\ \ddot{x}_{19}^d \\ \ddot{x}_{21}^d \end{bmatrix}, \bar{f}(x) = \begin{bmatrix} \bar{f}_{18}(x) \\ \bar{f}_{20}(x) \\ \bar{f}_{22}(x) \end{bmatrix},$$

$$\Lambda = \begin{bmatrix} \lambda_z & 0 & 0 \\ 0 & \lambda_\theta & 0 \\ 0 & 0 & \lambda_\phi \end{bmatrix}, \dot{e} = \begin{bmatrix} \dot{e}_z \\ \dot{e}_\theta \\ \dot{e}_\phi \end{bmatrix}, k = \begin{bmatrix} k_z & 0 & 0 \\ 0 & k_\theta & 0 \\ 0 & 0 & k_\phi \end{bmatrix}, \text{and } s = \begin{bmatrix} s_z \\ s_\theta \\ s_\phi \end{bmatrix}.$$

Functions g_{18}, g_{20}, and g_{22} are nonzero scalars. Thus, they are invertible. So is G. Note that the constant rate reaching law $(Msign(s))$ is not used in (5.24). This is because the use of the proportional rate reaching law (ks) proved to be sufficient.

The three desired inputs u_z^d, u_θ^d, and u_ϕ^d can be interpreted as being, respectively, the desired force in the z direction and the two desired torques in the direction of angles θ and ϕ so that the controlled system reaches its desired objectives. However, the relation between u_1 and the actuator forces f is

$$u_1 = \begin{bmatrix} u_z \\ u_\theta \\ u_\phi \end{bmatrix} = \begin{bmatrix} 1 & 1 & 1 & 1 \\ a & a & -b & -b \\ -c & d & -c & d \end{bmatrix} \begin{bmatrix} f_{fr} \\ f_{fl} \\ f_{rr} \\ f_{rl} \end{bmatrix} = Nf. \tag{5.25}$$

Thus, for the desired u_1^d of (5.24) the four desired actuator forces are given by

$$f^d = N^+ u_1^d. \tag{5.26}$$

$N^+ = N^T(NN^T)^{-1}$ is the pseudo-inverse of matrix N. The pseudo-inverse of this matrix has an exact solution. That is to say that N is of full row rank $(rank(N) = 3)$ and is right invertible $(NN^+ = I_3)$. I_3 is the identity matrix of dimension three. The desired actuation forces f^d in (5.26) are determined by the *decoupling* module of the control scheme (Fig. 5.6).

The desired actuators pressures are obtained by dividing the desired forces (5.26) by the sectional area S of the actuator piston:

$$A^d = \begin{bmatrix} A_{fr}^d & A_{fl}^d & A_{rr}^d & A_{rl}^d \end{bmatrix}^T = \frac{1}{S} \begin{bmatrix} f_{fr}^d & f_{fl}^d & f_{rr}^d & f_{rl}^d \end{bmatrix}^T = \frac{1}{S} f^d. \tag{5.27}$$

Once the four desired actuators pressures A^d are determined by $C_{Chassis}$, the control modules C_ϑ^c and C_ϑ^s are designed to generate these desired pressures.

5.3.2 Second Control Module: Cylinder Module C_ϑ^c

The desired actuator pressure A_ϑ^d is determined by the chassis control module $C_{Chassis}$ and the *decoupling* module. The role of C_ϑ^c is to determine the spool valve position $z_{v_\vartheta}^d$ needed to generate the desired actuator pressure A_ϑ^d (see Fig. 5.6). The pressure generated by an actuator is given by

$$\dot{A}_\vartheta = -\beta A_\vartheta - \alpha S(\dot{z}_{s_\vartheta} - \dot{z}_{u_\vartheta}) + sign(p_s - sign(z_{v_\vartheta})A_\vartheta)\gamma\sqrt{|p_s - sign(z_{v_\vartheta})A_\vartheta|}z_{v_\vartheta}. \tag{5.28}$$

This notation is used in place of the state-space representation to generalize the study of the second module for the four force actuators. This equation can be written as

$$\dot{A}_\vartheta = f_{2_\vartheta} + g_{2_\vartheta} u_{2_\vartheta}, \tag{5.29}$$

where $f_{2_\vartheta} = -\beta A_\vartheta - \alpha S(\dot{z}_{s_\vartheta} - \dot{z}_{u_\vartheta})$,
$u_{2_\vartheta} = z_{v_\vartheta}$ and $g_{2_\vartheta} = sign(p_s - sign(z_{v_\vartheta})A_\vartheta)\gamma\sqrt{|p_s - sign(z_{v_\vartheta})A_\vartheta|}$.

Defining the sliding surface as

$$s_{2_\vartheta} = A_\vartheta - A_\vartheta^d, \tag{5.30}$$

and following the steps described in Sect. 5.3, the second control law can be given by [2]

$$u_{2_\vartheta}^d = z_{v_\vartheta}^d = \frac{\beta A_\vartheta + \alpha S(\dot{z}_{s_\vartheta} - \dot{z}_{u_\vartheta}) + \dot{A}_\vartheta^d - k_{2_\vartheta} s_{2_\vartheta}}{sign(p_s - sign(z_{v_\vartheta})A_\vartheta)\gamma\sqrt{|p_s - sign(z_{v_\vartheta})A_\vartheta|}}. \tag{5.31}$$

k_{2_ϑ} is chosen to satisfy (2.58) with $\vartheta \in \{fr, \ fl, \ rr, \ rl\}$.
The vector of the four desired spool valve positions is

$$z_v^d = \begin{bmatrix} u_{2_{fr}}^d & u_{2_{fl}}^d & u_{2_{rr}}^d & u_{2_{rl}}^d \end{bmatrix}^T = \begin{bmatrix} z_{v_{fr}}^d & z_{v_{fl}}^d & z_{v_{rr}}^d & z_{v_{rl}}^d \end{bmatrix}^T. \tag{5.32}$$

5.3.3 Third Control Module: Servo Valve Module C_ϑ^s

The required spool valve position $z_{v_\vartheta}^d$ to generate the desired actuator pressure A_ϑ^d (determined by $C_{Chassis}$) is computed by C_ϑ^s. The role of the control module C_ϑ^s is to determine the control input u_ϑ needed to obtain $z_{v_\vartheta}^d$. The spool valve dynamics are

$$\dot{z}_{v_\vartheta} = \frac{1}{\tau}(-z_{v_\vartheta} + u_\vartheta). \tag{5.33}$$

These dynamics can be written as

$$\dot{z}_{v_\vartheta} = f_{3_\vartheta} + g_{3_\vartheta} u_{3_\vartheta}, \tag{5.34}$$

with $f_{3_\vartheta} = -\frac{z_{v_\vartheta}}{\tau}$, $g_{3_\vartheta} = \frac{1}{\tau}$ and $u_{3_\vartheta} = u_\vartheta$. Defining the sliding surface as

$$s_{3_\vartheta} = z_{v_\vartheta} - z_{v_\vartheta}^d, \tag{5.35}$$

and following the steps described in Sect. 5.3, the third control law can be given by [2]

$$u_{3_\vartheta}^d = u_\vartheta^d = z_{v_\vartheta} + \tau(\dot{z}_{v_\vartheta}^d - k_{3_\vartheta} s_{3_\vartheta}). \tag{5.36}$$

k_{3_ϑ} is chosen to satisfy (2.58) with $\vartheta \in \{fr,\ fl,\ rr,\ rl\}$. The control input u is expressed as follows:

$$u = \begin{bmatrix} u_{fr}^d & u_{fl}^d & u_{rr}^d & u_{rl}^d \end{bmatrix}^T. \tag{5.37}$$

5.4 System Instrumentation

The control law requires an investigation into the needed measurements and the available and the unavailable sensors. Therefore, this section will analyze the problem of active suspension instrumentation.

5.4.1 Required Measurements

The control law requires the measurement/estimation of:

- The heave position z
- The heave velocity $\dot{z}$
- The pitch angle θ
- The pitch angular velocity $\dot{\theta}$
- The roll angle ϕ
- The roll angular velocity $\dot{\phi}$
- The four suspensions stroke $z_{s_\vartheta} - z_{u_\vartheta}$
- The four suspensions stroke rate $\dot{z}_{s_\vartheta} - \dot{z}_{u_\vartheta}$
- The four spool valve positions z_{v_ϑ}

- The four actuator forces f_ϑ with $\vartheta \in \{fr, \ fl, \ rr, \ rl\}$

Note that in the system model the suspension stroke $z_{s_\vartheta} - z_{u_\vartheta}$ and stroke rate $\dot{z}_{s_\vartheta} - \dot{z}_{u_\vartheta}$ are expressed in function of the state components using the physical relations between the chassis (z, θ and ϕ) and its four corners z_{s_ϑ}. These relations are

$$\begin{cases} z_{s_{fr}} = x_{17} + a\sin(x_{19}) - c\sin(x_{21}) \\ z_{s_{fl}} = x_{17} + a\sin(x_{19}) + d\sin(x_{21}) \\ z_{s_{rr}} = x_{17} - b\sin(x_{19}) - c\sin(x_{21}) \\ z_{s_{rl}} = x_{17} - b\sin(x_{19}) + d\sin(x_{21}) \end{cases} . \tag{5.38}$$

The required measurements problem needs an investigation into the available sensors to provide these measurements.

5.4.2 Available Sensors

The following study details two working cases; the *prototype* and the *industrial* levels.

Prototype Vehicles

The prototype vehicles are used in the laboratories for research purposes. The most common sensors that can be found are the following:

- Laser sensors measure the heave position z of the sprung mass, the displacement z_{u_ϑ} of the unsprung masses and the road inputs z_{r_ϑ} ($\vartheta \in \{fr, \ fl, \ rr, \ rl\}$)
- Accelerometers measure the heave acceleration $\ddot{z}$ of the sprung mass and the vertical acceleration $\ddot{z}_{u_\vartheta}$ of the unsprung masses
- Accelerometers measure the lateral and longitudinal acceleration of the vehicle
- Gyrometers measure pitch and roll angular velocities $\dot{\theta}$ and $\dot{\phi}$
- Linear variable displacement transducers measure the suspensions stroke $z_{s_\vartheta} - z_{u_\vartheta}$, the sprung mass displacement z_{u_ϑ}, the road displacement z_{r_ϑ}, and the spool valve positions z_{v_ϑ}
- Linear velocity transducers measure the sprung mass vertical velocity $\dot{z}$, the suspensions stroke rate $\dot{z}_{s_\vartheta} - \dot{z}_{u_\vartheta}$, and the unsprung masses vertical velocities $\dot{z}_{u_\vartheta}$
- Load cells measure the actuators forces f_ϑ

Industrial Vehicles

Not all the sensors cited above are on board. The sensors combination depends mainly on the control law and the sensors costs. The most common sensors available are

- The laser sensors
- The accelerometers
- The gyrometers
- The linear variable displacement transducers which measure the suspensions stroke $z_{s_\vartheta} - z_{u_\vartheta}$

Here are some examples of the sensors used in standard vehicles:

- The Maserati GranSport uses four acceleration sensors on each wheel (skyhook system)
- The 2005 Cadillac XLR world's fastest-reacting suspension system is equipped with four wheel-to-body displacement sensor that measure wheel motion over the road surface
- The Opel Astra uses five acceleration sensors: three body sensors that measure roll, pitch, and heave and two wheel sensors for the road surface conditions
- The Volvo S60 R and V70 R, equipped with the Four-C technology (continuously controlled chassis concept), uses sensors measuring longitudinal and lateral acceleration, yaw rates, roll, pitch and heave, vertical position of each wheel, speed of the car, steering wheel position and velocity, engine torque, throttle pedal position, engine rpm, and braking force

5.4.3 Unavailable Sensors

System states that are not measurable are

- The heave velocity $\dot{z}$: sensors allowing the measurement of the sprung mass heave velocity $\dot{z}$ do not exist. On prototype vehicles, this measurement is ensured using a linear velocity transducer and a fixed reference (*i.e.*, the ground) which is not possible to realize on industrial vehicles. Nevertheless, two solutions are possible. The first uses a laser sensor to measure z and then a smooth derivative technique gives $\dot{z}$. The second solution uses an accelerometer for the heave acceleration $\ddot{z}$ measurement and an integration leads to $\dot{z}$.
- The pitch θ and roll ϕ angles: sensors measuring the inclination angles do exist. However, these sensors are sensitive to vibration. Sensors with high vibration resistance are very expensive and are still inadequate to be used in vibrating environments such as vehicles. Therefore, gyrometers are used to measure $\dot{\theta}$ and $\dot{\phi}$. Angles are then obtained by integrating the measurements.

5.4.4 Sensors in Use

For the SMC of the nonlinear system, 15 sensors are assumed to be used:

- Five accelerometers measure the heave acceleration $\ddot{z}$ and the four unsprung masses acceleration $\ddot{z}_{u_\vartheta}$. The noisy measurements are filtered [68]. Velocities and positions are obtained by double integration [68].
- Two gyrometers measure the pitch θ and roll ϕ angular velocities. Integration leads to the respective angles [68].
- Four linear variable displacement transducers measure the spool valve positions z_{v_ϑ} [2].
- Four load cells measure the actuators forces f_ϑ [2].

The four suspension stroke rates $\dot{z}_{s_\vartheta} - \dot{z}_{u_\vartheta}$ are computed using the physical relations (5.38) between the chassis and its four corners z_{s_ϑ} ($\vartheta \in \{fr, fl, rr, rl\}$).

This sensor combination represents one of many possible sensor solutions. However, the one proposed here is close to those used for industrial vehicles. Thus, it respects the industrial constraints.

Obtaining velocities and positions by integrating acceleration measurements is used in [68] for a prototype vehicle. A filtering of the noisy measurement is performed before the integration. Integrating the measurement is inevitable in obtaining the variables needed for the control. However, this may lead to bias over a certain time window. This affects the control and the diagnosis scheme. This point is not investigated in this chapter. This is considered as an area of future research.

Remark 5.1. The used sensors will be referred to as follows:

- For $SS_{chassis}$, $\zeta^1_{chassis}$, $\zeta^2_{chassis}$, and $\zeta^3_{chassis}$ denote the three sensors that measure $\ddot{z}$, $\dot{\theta}$, and $\dot{\phi}$ respectively
- For SS_ϑ, ζ^1_ϑ, ζ^2_ϑ, and ζ^3_ϑ denote the three sensors that measure $\ddot{z}_{u_\vartheta}$, f_ϑ, and z_{v_ϑ} ($\vartheta \in \{fr,\ fl,\ rr,\ rl\}$)

5.4.5 Optimal Sensor Network Design

In this study, 15 sensors are assumed to be used. For cost reasons, not all of these sensors may be installed on standard vehicles. However, some material redundancy should be available for the diagnosis and fault accommodation. Thus, the minimal number of sensors ensuring the observability of the system and allowing the FDI must be found. This point is treated in [19, 20] and will not be presented here.

5.5 Sensor Fault Diagnosis Strategy

Like any controlled system, the behavior of the active suspension system depends on, among other things, the control scheme and the information delivered by the sensors. Thus, any incorrect information caused by a faulty sensor can lead the system to an undesirable or dangerous behavior.

Few research studies have treated the suspension system fault diagnosis problem. An early study used the generalized likelihood ratio approach [88] for on-board sensor FDI of an active suspension system using a half-car model. It was concluded that the application of this approach is feasible when the failure can be modeled as a deterministic additive term. In other situations, the computational requirements make it less practical. Another study uses the nearest neighbor-based fault identification and the robust geometric methodology [107].

Isermann and other researchers of the Darmstadt University of Technology are working on this problem: model-based methods are applied to a quarter car test rig equipped with an electro-hydraulic actuator [34–40, 72]. In these works, the unknown parameters of the system are estimated by using an estimation algorithm. Then parity equations are used for the detection and isolation of sensors and components faults. Parity equations are generated with a local linear model tree.

Other methods are used for the FDI in active suspension: statistical methodologies are applied to perform fault detection in nonlinear two degrees of freedom quarter-car model and complete vehicle models [96]. Model-based FDI methods are developed in [15].

In [74], analytical redundancy techniques are applied to fault detection for heavy vehicles where the full vehicle model is divided into several subsystems to reduce the computation cost. In [77], the model-based fault detection approach relies on simple mathematical descriptions of the system which yields a robust FDI against disturbances or model uncertainties.

In this section, a model-based sensor FDI strategy is designed for the full vehicle active suspension system. The consideration of the force actuators helps mainly in better understanding the occurrence of the faults and their propagation in the whole system. This is because the actuators are crucial elements of the active suspension systems. They have their proper dynamics and they are also subject to faults. Six diagnosis modules are then designed: a global one localizes the fault in one of the subsystems and five local diagnosis modules (one for each subsystem) determine the faulty sensor (see Figs. 5.4 and 5.5).

To detect and isolate sensor failures, the system model and a bank of SMOs are used to generate residuals. In this study, abrupt faults are considered. The residuals are designed in such a way that every fault has a specific pattern and thus can be easily isolated.

5.5.1 Sliding Mode Observers-based Sensor Fault Diagnosis

This section shows how SMO can be used for detecting faults of large magnitude. This SMO-based FDI scheme is not useful for detecting incipient faults in many practical applications where system model mismatch is unavoidable and significant [17]. An incipient fault will neither produce a sudden peak in the residual nor will it push the system out of sliding mode. This is a limitation of this SMO-based fault diagnosis.

The subsequent explanation of the SMO-based FDI scheme is derived from [17]. The initial scheme is reformulated in order to be compatible with the use of reduced order observers. For this purpose, consider a reduced order multivariable nonlinear system described in state-space form:

$$\begin{cases} \dot{x} = f(x, \tilde{x}, u, d) \\ y = H(x) + f_o(t) \end{cases},$$ (5.39)

where $x \in M$, a C^∞ connected manifold of dimension n. The use of reduced order systems consists in considering a part of the system model. Thus $\tilde{x}$ are the states other than x. $\tilde{x}$ are considered as known inputs. u and d represent respectively the control input and the system uncertainties. It is assumed that all x, $\tilde{x}$, u, and d are bounded. $H(x) = [h_1(x), \cdots, h_p(x)]^T$ are smooth vector fields on M. $f_o(t) = [f_o^1(t), \cdots, f_o^p(t)]^T$ represents the sensor faults.

A general SMO for (5.39) is of the following form:

$$\dot{z} = L(\tilde{x} + f_s(t), z, y, u) + \Lambda sign(y - z),$$ (5.40)

where z is an estimate of $H(x)$. L is a function of $\tilde{x}$, z, y, and u. Λ is a diagonal gain matrix with elements λ_i, $i = 1, \cdots, p$. The term $f_s(t)$ represents the faults of the sensors measuring $\tilde{x}$. It can be any function of time.

The estimation error is given by $e = H(x) - z$. The dynamics of the estimation error are given by

$$\dot{e} = \frac{d}{dt} H(x) - \dot{z} = \frac{\partial H(x)}{\partial x} \dot{x} - \dot{z}$$ (5.41)

$$= \frac{\partial H(x)}{\partial x} f(x, \tilde{x}, u, d) - L(\tilde{x} + f_s(t), z, y, u) - \Lambda sign(H(x) + f_o(t) - z).$$

Note that error e cannot be measured and that only $r = y - z = H(x) + f_o(t) - z$ is measurable. The quantity r is considered as being the sliding surface; thus it can be considered as a residual signal.

Assume that $e_i = h_i(x) - z_i$ and define

$$\frac{\partial H(x)}{\partial x} f(x, \tilde{x}, u, d) = \begin{bmatrix} m_1(x, \tilde{x}, u, d) \\ m_2(x, \tilde{x}, u, d) \\ \vdots \\ m_p(x, \tilde{x}, u, d) \end{bmatrix} \text{ and } L = \begin{bmatrix} l_1(\tilde{x} + f_s(t), z, y, u) \\ l_2(\tilde{x} + f_s(t), z, y, u) \\ \vdots \\ l_p(\tilde{x} + f_s(t), z, y, u) \end{bmatrix}.$$

The dynamics of the error e_i are then given by

$$\dot{e}_i = m_i(x, \tilde{x}, u, d) - l_i(\tilde{x} + f_s(t), z, y, u) - \lambda_i sign(e_i + f_o^i(t)).$$ (5.42)

When no sensor fault is present, $f_s(t) = f_o(t) = 0$. If λ_i is chosen such that $|m_i(x, \tilde{x}, u, d) - l_i(\tilde{x}, z, H(x), u)| \le \lambda_i$:

$$\frac{d}{dt}e_i^2 = 2e_i[m_i(x, \tilde{x}, u, d) - l_i(\tilde{x}, z, H(x), u) - \lambda_i sign(e_i)]. \tag{5.43}$$

Therefore:

if $e_i > 0$, $\frac{d}{dt}e_i^2 = 2e_i\dot{e}_i = 2e_i[m_i(x, \tilde{x}, u, d) - l_i(\tilde{x}, z, H(x), u) - \lambda_i] < 0$,

if $e_i < 0$, $\frac{d}{dt}e_i^2 = 2e_i\dot{e}_i = 2e_i[m_i(x, \tilde{x}, u, d) - l_i(\tilde{x}, z, H(x), u) + \lambda_i] < 0$.

Thus $r_i = e_i$ exponentially decreases to zero according to the Lyapunov principle [17]. Assume now that one of the sensors measuring $\tilde{x}$ becomes faulty at time t_i ($f_s(t) \neq 0$). It is necessary to investigate how the fault interacts with the sliding surface and how the sliding performance of the observer is affected.

At time $t > t_i$, $f_s(t) \neq 0$. Then

$$\dot{r}_i = m_i(x, \tilde{x}, u, d) - l_i(\tilde{x} + f_s(t), z, y, u) - \lambda_i sign(e_i). \tag{5.44}$$

This may produce one of the following two cases.

Case 1

If $|m_i(x, \tilde{x}, u, d) - l_i(\tilde{x} + f_s(t), z, y, u)| \leq \lambda_i$ for all $i = 1, \cdots, p$, the observer will remain on the sliding surface and faults can not be detected, i.e., $r_i = 0$, $i = 1, \cdots, p$.

Case 2

If $|m_i(x, \tilde{x}, u, d) - l_i(\tilde{x} + f_s(t), z, y, u)| \geq \lambda_i$ persistently holds, the observer will move from its surface and sliding will cease. In this case, the ith residual element will become nonzero persistently and will alarm for the occurrence of a sensor fault.

The effect of sensor faults $f_o(t)$ is also considered in the FDI design but it is not presented here. The idea for these faults is similar to that of $f_s(t)$. More details about this subject are given in [17].

5.5.2 Active Suspension Sensor Fault Detection and Isolation

Sensors are generally exposed to noise, outliers, gains, offsets, breakdown (ground), or freeze at the minimum or the maximum value. However, this study considers abrupt sensor faults like gains, offsets, breakdown, and freeze. Other fault types will be considered in future works.

A model-based approach is used to detect a sensor fault and to isolate it. The system model and a bank of SMOs generate residuals. The residuals are designed in such a way that each possible fault has a unique pattern. They define *a priori* known fault-signature inference matrices. In total, six inference matrices are used to isolate faults. One is a global matrix for the whole system. The others are local matrices, one for each subsystem. The design of these matrices is detailed in the subsequent sections.

Diagnosis Strategy

The fault diagnosis strategy is illustrated in Fig. 5.7. It consists of six diagnosis modules: one global and five local $Diag_\vartheta$ for each subsystem SS_ϑ with $\vartheta \in \{fr, fl, rr, rl, chassis\}$.

To construct the global diagnosis module, the system model is used to estimate, with a geometric approach, the actuation pressures $\hat{x}_{4.k-1}$ of the four actuators $(k = 1, \cdots, 4)$. Here, the "." represent multiplication. The estimated pressures $\hat{x}_{4.k-1}$ are then compared to the measurements to generate the four residuals R_k $(k = 1, \cdots, 4)$.

For the local diagnosis modules $Diag_\vartheta$, six reduced order SMO $R.Obs.i$ $(i = 1, \cdots, 6)$ estimate the heave velocity of the sprung mass $(x_{18})_k$ $(k = 5, \cdots, 10)$. The estimates are then compared to the measurement of the heave velocity x_{18} to generate six additional residuals R_k $(k = 5, \cdots, 10)$. The fault isolation is ensured by injecting the output vector y in each observer while excluding some measurements. The notation "$\backslash m$" in Fig. 5.7 denotes that the measurement "m" is excluded from the output vector y. This helps in decoupling the effect of the excluded measurement from the estimated variables.

Finally, ten estimates are obtained and compared to the measurements. Ten residuals R_k $(k = 1, \cdots, 10)$ are generated to detect and isolate sensor faults. The global and the five local modules $Diag_\vartheta$ are detailed in the sequel with $\vartheta \in \{fr, fl, rr, rl, chassis\}$.

Global Fault Diagnosis Module

To design the global fault diagnosis module, the system model is used to define four residuals and a global inference matrix. From $\dot{x}_{4.k-1}$ $(k = 1, \cdots, 4)$, the estimate $\dot{\hat{x}}_{4.k-1}$ is calculated as function of $x_{4.k-2}$, $x_{4.k-1}$, $x_{4.k}$, x_{18}, x_{19}, x_{20}, x_{21}, and x_{22} as follows:

$$\dot{\hat{x}}_{4.k-1} = f(x_{4.k-2}, x_{4.k-1}, x_{4.k}, x_{18}, x_{19}, x_{20}, x_{21}, x_{22}) + h_k sign(R_k), \quad (5.45)$$

where $R_k = x_{4.k-1} - \hat{x}_{4.k-1}$. $x_{4.k-1}$ is the measured actuator load pressure and $\hat{x}_{4.k-1}$ is the

$$k^{\text{th}}$$

calculated or estimated one. h_k is a positive constant to be chosen and $k = 1, \cdots, 4$.

In the ideal case where the model is perfect, residual R_k nominally equals zero. When any sensor measuring $x_{4.k-2}$, $x_{4.k-1}$, $x_{4.k}$, x_{18}, x_{19}, x_{20}, x_{21}, or x_{22} is faulty, $\hat{x}_{4.k-1}$ deviates from $x_{4.k-1}$ and R_k deviates from zero $(k = 1, \cdots, 4)$.

Remark 5.2. In fact, no sensors measure $x_{4.k-3}$, $x_{4.k-2}$, x_{17}, x_{18}, x_{19}, or x_{21} $(k = 1, \cdots, 4)$. These variables are obtained by integrating the measurements (see Sect. 5.4.4):

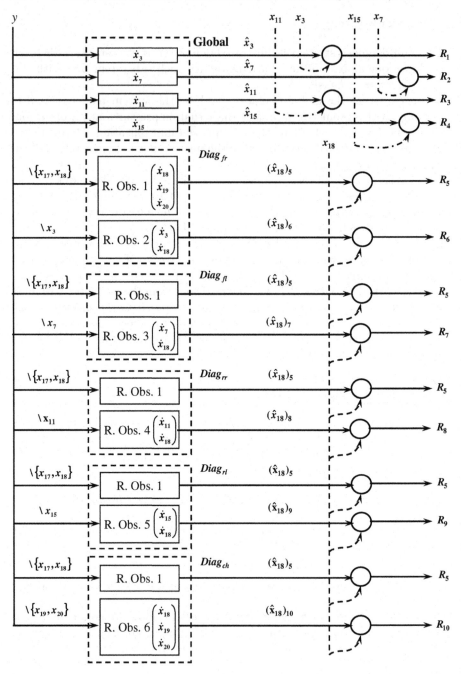

Fig. 5.7. Scheme of the diagnosis strategy

- Sensor $\zeta_{chassis}^1$ measures $\ddot{z}$; the heave z (x_{17}) and the heave velocity $\dot{z}$ (x_{18}) are obtained by integrating $\ddot{z}$
- Sensor $\zeta_{chassis}^2$ measures the pitch angular velocity $\dot{\theta}$ (x_{20}); the pitch angle θ (x_{19}) is obtained by integrating this measurement
- Sensor $\zeta_{chassis}^3$ measures the roll angular velocity $\dot{\phi}$ (x_{22}); the roll angle ϕ (x_{21}) is obtained by integrating this measurement
- Sensor ζ_{ϑ}^1 measures $\ddot{z}_{u_\vartheta}$ ($\vartheta \in \{fr, fl, rr, rl\}$); the unsprung mass displacement z_{u_ϑ} ($x_{4.k-3}$) and the unsprung mass velocity $\dot{z}_{u_\vartheta}$ ($x_{4.k-2}$) are obtained by integrating the measurement ($k = 1, \cdots, 4$)
- Sensor ζ_{ϑ}^2 measures f_ϑ ($x_{4.k-1}$) ($k = 1, \cdots, 4$)
- Sensor ζ_{ϑ}^3 measures z_{v_ϑ} ($x_{4.k}$) ($k = 1, \cdots, 4$)

A sensor fault results in an error in the measurement and the variables derived from it.

With these four defined residuals a fault-signature matrix is obtained in Table 5.2. The "1" means that the residual is sensitive to the fault occurrence, while the "0" means that it is insensitive.

Table 5.2. Global fault-signature matrix

	SS_{fr}			SS_{fl}			SS_{rr}			SS_{rl}			SS_{ch}		
	ζ_{fr}^1	ζ_{fr}^2	ζ_{fr}^3	ζ_{fl}^1	ζ_{fl}^2	ζ_{fl}^3	ζ_{rr}^1	ζ_{rr}^2	ζ_{rr}^3	ζ_{rl}^1	ζ_{rl}^2	ζ_{rl}^3	ζ_{ch}^1	ζ_{ch}^2	ζ_{ch}^3
R_1	1	1	1	0	0	0	0	0	0	0	0	0	1	1	1
R_2	0	0	0	1	1	1	0	0	0	0	0	0	1	1	1
R_3	0	0	0	0	0	0	1	1	1	0	0	0	1	1	1
R_4	0	0	0	0	0	0	0	0	0	1	1	1	1	1	1

This inference matrix makes it possible to detect the occurrence of a sensor fault and to locate it in one of the five subsystems. This is mainly done by the global FDI module which supervises the whole system. For instance, if the sensor ζ_{fr}^2 measuring x_3 breaks down, only the residual R_1 is affected. The fault is detected but not isolated. The global FDI module indicates that the fault is present in the front right subsystem (SS_{fr}) since it has the signature $(1, 0, 0, 0)^T$, without specifying the defected sensor. Complementary investigation should be performed by the local FDI module $Diag_{fr}$ of (SS_{fr}) to determine the faulty sensor.

In the sequel, the local FDI modules $Diag_\vartheta$ ($\vartheta \in \{fr, fl, rr, rl, chassis\}$) are presented. These modules are designed using a bank of SMO.

Front Right Subsystem Diagnosis Module $Diag_{fr}$

To isolate faults in SS_{fr}, two reduced order SMO are built. A reduced order observer is constructed with part of the system model. The consideration of part of the model rather than the complete model reduces computation cost and simplifies observer design. This part must be observable.

The first observer R.Obs.1 is constructed using the set of equations $\dot{x}_{18}$, $\dot{x}_{19}$, and $\dot{x}_{20}$. The variables x_{17} and x_{18} are assumed to be non-measured (they are excluded from the output vector y (Fig. 5.7)). This observer estimates x_{18}. The considered equations are

$$\dot{x}_{18} = \{k_{s_{fr}}x_1 + k_{s_{fl}}x_5 + k_{s_{rr}}x_9 + k_{s_{rl}}x_{13} - (k_{s_{fr}} + k_{s_{fl}} + k_{s_{rr}} + k_{s_{rl}})x_{17} - [a(k_{s_{fr}}+k_{s_{fl}})-b(k_{s_{rr}}+k_{s_{rl}})]\sin(x_{19}) - [d(k_{s_{fl}}+k_{s_{rl}})-c(k_{s_{fr}}+k_{s_{rr}})]\sin(x_{21}) + c_{s_{fr}}x_2 + c_{s_{fl}}x_6 + c_{s_{rr}}x_{10} + c_{s_{rl}}x_{14} - (c_{s_{fr}} + c_{s_{fl}} + c_{s_{rr}} + c_{s_{rl}})x_{18} - [a(c_{s_{fr}} + c_{s_{fl}}) - b(c_{s_{rr}} + c_{s_{rl}})]\cos(x_{19})x_{20} - [d(c_{s_{fl}}+c_{s_{rl}}) - c(c_{s_{fr}}+c_{s_{rr}})]\cos(x_{21})x_{22} + S(x_3 + x_7 + x_{11} + x_{15})\}/M,$$

$$\dot{x}_{19} = x_{20},$$

$$\dot{x}_{20} = \cos(x_{19})\{ak_{s_{fr}}x_1 + ak_{s_{fl}}x_5 - bk_{s_{rr}}x_9 - bk_{s_{rl}}x_{13} - [a(k_{s_{fr}}+k_{s_{fl}})-b(k_{s_{rr}}+k_{s_{rl}})]x_{17} - [a^2(k_{s_{fr}}+k_{s_{fl}})+b^2(k_{s_{rr}}+k_{s_{rl}})]\sin(x_{19}) - [d(ak_{s_{fl}}-bk_{s_{rl}})-c(ak_{s_{fr}}-bk_{s_{rr}})]\sin(x_{21}) + ac_{s_{fr}}x_2 + ac_{s_{fl}}x_6 - bc_{s_{rr}}x_{10} - bc_{s_{rl}}x_{14} - [a(c_{s_{fr}}+c_{s_{fl}})-b(c_{s_{rr}}+c_{s_{rl}})]x_{18} - [a^2(c_{s_{fr}} + c_{s_{fl}}) + b^2(c_{s_{rr}} + c_{s_{rl}})]\cos(x_{19})x_{20} - [d(ac_{s_{fl}} - bc_{s_{rl}}) - c(ac_{s_{fr}} - bc_{s_{rr}})]\cos(x_{21})x_{22} + S[a(x_3 + x_7) - b(x_{11} + x_{15})]\}/I_{yy}.$$

According to these equations, the estimation of x_{18} is explicitly independent of x_4. Thus, the estimate $(\hat{x}_{18})_5$ and the residual R_5 are insensitive to the fault of the sensor ζ_{fr}^3 that measures the front-right spool valve position x_4. On the other hand, they are sensitive to the other measurements.

Remark 5.3. The estimate $(\hat{x}_{18})_5$ obtained from R.Obs.1 is independent of x_8, x_{12}, and x_{16}. Thus, it will be used for the generation of residuals in the other subsystems.

The second observer R.Obs.2 is designed using $\dot{x}_3$ and $\dot{x}_{18}$. The variable x_3 is assumed to be non-measured (it is excluded from the output vector y as illustrated in Fig. 5.7). Thus, $(\hat{x}_{18})_6$ and R_6 are insensitive to the measurement of the front-right actuator load pressure x_3. They are sensitive to the other measurements x_{18} and $\tilde{x}$.

Two additional residuals $R_k = x_{18} - (\hat{x}_{18})_k$ are then obtained, with $k = \{5, 6\}$. x_{18} is the measured heave velocity and $(\hat{x}_{18})_k$ is the

$$k^{\text{th}}$$

estimated one.

For this subsystem the inference matrix is given in Table 5.3 with $\vartheta = fr$ and $l = 0$. With both additional residuals, it is possible to isolate the faults of this subsystem. The same steps are made for the other subsystems.

Table 5.3. Local faults-signature inference matrix for SS_ϑ

	ζ_ϑ^1	ζ_ϑ^2	ζ_ϑ^3
R_5	1	1	0
R_{6+l}	1	0	1

Front Left Subsystem Diagnosis Module $Diag_{fl}$

This module has $R.Obs.1$ as the first observer (see Remark 5.3). This observer is insensitive to the fault of the sensor ζ_{fl}^3 measuring the front-left spool valve position x_8 and sensitive to the other measurements.

The second observer $R.Obs.3$ is designed using $\dot{x}_7$ and $\dot{x}_{18}$. The variable x_7 is not assumed to be measured (it is excluded from the output vector y as illustrated in Fig. 5.7). Thus, $(\hat{x}_{18})_7$ and R_7 are insensitive to the fault of the sensor ζ_{fl}^2 measuring the front-left actuator load pressure x_7. They are sensitive to the other measurements x_{18} and $\tilde{x}$.

Two residuals $R_k = x_{18} - (\hat{x}_{18})_k$ are then obtained, with $k = \{5, 7\}$. Table 5.3 illustrates the inference matrix of this subsystem with $\vartheta = fl$ and $l = 1$.

Rear Right Subsystem Diagnosis Module $Diag_{rr}$

The first observer for this module is $R.Obs.1$ (see Remark 5.3) that is insensitive to the fault of sensor ζ_{rr}^3 measuring the rear-right spool valve position x_{12}.

For the two equations $\dot{x}_{11}$ and $\dot{x}_{18}$, the observer $R.Obs.4$ is designed with the assumption that the variable x_{11} is not measured. Thus $(\hat{x}_{18})_8$ and R_8 are insensitive to the fault of sensor ζ_{rr}^2 measuring the rear-right actuator load pressure x_{11}. They are sensitive to the other measurements.

Two residuals $R_k = x_{18} - (\hat{x}_{18})_k$ are then obtained, with $k = \{5, 8\}$. The inference matrix for this subsystem is given in Table 5.3 with $\vartheta = rr$ and $l = 2$.

Rear Left Subsystem Diagnosis Module $Diag_{rl}$

In addition to the observer $R.Obs.1$ being insensitive to the fault of ζ_{rl}^3, a second observer $R.Obs.5$ is designed for the two equations $\dot{x}_{15}$ and $\dot{x}_{18}$, with the assumption that the variable x_{15} is not measured. Two residuals $R_k = x_{18} - (\hat{x}_{18})_k$ are obtained, with $k = \{5, 9\}$. The inference matrix for this subsystem is given in Table 5.3 with $\vartheta = rl$ and $l = 3$.

Chassis Subsystem Diagnosis Module $Diag_{ch}$

As for the preceding diagnosis modules, this module has a first observer $R.Obs.1$ that is sensitive to the faults of $\zeta_{chassis}^1$ and $\zeta_{chassis}^2$ and insensitive to that of $\zeta_{chassis}^3$.

The second observer $R.Obs.6$ is constructed for the three equations $\dot{x}_{18}$, $\dot{x}_{19}$ and $\dot{x}_{20}$, with the assumption that variables x_{19} and x_{20} are not measured. The two additional residuals $R_k = x_{18} - (\hat{x}_{18})_k$ are obtained, with $k = \{5, 10\}$. The inference matrix for this subsystem is given in Table 5.4.

Table 5.4. Chassis subsystem inference matrix

	$\zeta_{chassis}^1$	$\zeta_{chassis}^2$	$\zeta_{chassis}^3$
R_5	1	1	0
R_{10}	1	0	0

In conclusion, with these five local diagnosis modules, the sensor faults can be detected and the faulty sensors can be isolated.

5.5.3 Sensor Fault-tolerance

Once the fault is detected and isolated, the next step consists of taking a decision to reduce its effect on the system behavior. Very few studies treat the suspension fault-tolerance problem: a design of a bank of Kalman filters, one for each possible sensor failure configuration, providing an estimate of the system state when a sensor fault occurs, is carried out for a quarter car test rig [113]. A methodology for the comparison among different alternative fault-tolerant architectures, based on risk evaluation, has been applied to a full active suspension control system [16].

Sensor faults can be accommodated by replacing the faulty measurement by its estimation until the faulty sensor is replaced. This is known as sensor masking. In order to exploit their robustness to variation in system parameters, external disturbances, and modeling errors, a bank of SMO [132] is designed to provide an estimation of the different variables. Thus, it can be ready to be injected in the controller to replace the faulty measurement. The dynamics of such observers are given by (5.40).

However, the system observability may be lost for some sensor faults. The observability in the presence of a fault is investigated off-line. The fault-tolerant control system "knows" *a priori* when the system observability is preserved and when it is not. After FDI, the FTC system checks for the possibility of estimating the lost measurement. If possible, the lost measurement is estimated and replaced by its estimate. If not, the fault is considered to be critical and the system should be stopped safely.

Table 5.5 summarizes the possibility to accommodate sensor faults. The "1" means that the sensor fault can be compensated, the "0" means that it is not possible to estimate the erroneous measurement and that the fault is critical.

Table 5.5. Possibility of sensor fault-tolerance

	SS_{fr}			SS_{fl}			SS_{rr}			SS_{rl}			SS_{ch}		
	ζ_{fr}^1	ζ_{fr}^2	ζ_{fr}^3	ζ_{fl}^1	ζ_{fl}^2	ζ_{fl}^3	ζ_{rr}^1	ζ_{rr}^2	ζ_{rr}^3	ζ_{rl}^1	ζ_{rl}^2	ζ_{rl}^3	ζ_{ch}^1	ζ_{ch}^2	ζ_{ch}^3
Possibility	0	1	1	0	1	1	0	1	1	0	1	1	1	1	1

Fault-tolerance requires the coordination between the two structures of diagnosis and control. This coordination is performed with the module *decision taking and resources management* as illustrated in Fig. 5.5. This coordination is necessary for the reconfiguration of control structure in function of the remaining resources after fault occurrence.

5.6 Simulation

To illustrate the control and FDI approaches, simulation is made using MATLAB®/Simulink® with the solver ODE1 and a fixed step equals to 10^{-4}. In order to make it more realistic, white noise is added to measurements. The physical limits of the system and the actuator constraints are taken into consideration.

5.6.1 Frequency Response

For the linearized model, Fig. 5.8 illustrates the magnitude of the loop transfer $\left|\frac{\ddot{z}}{z_{r_{fl}}}\right|$ with the control law given by (5.24). The solid line represents the passive suspension and the dotted line represents the active suspension. $z_{r_{fl}}$ is the front left wheel disturbance. As shown in Fig. 5.8, the system acceleration is highly damped. However, this is difficult to realize in practice because of the physical limits of the actuators: the forces generated by the actuators are bounded and the actuator response is not fast enough to insure the damping of the road high frequencies perturbation.

If (5.24) is replaced by the following control law:

$$u_1^d = G^{-1}(\ddot{x}^d - \Lambda\dot{e} - ks), \tag{5.46}$$

then control input u_z given by (5.21) becomes

$$u_z = g_{18}^{-1}(\ddot{x}_{17}^d - \lambda_z\dot{e}_z - k_z s_z). \tag{5.47}$$

Substituting (5.47) in (5.18) gives

$$\dot{s}_z = \bar{f}_{18}(x) - k_z s_z, \tag{5.48}$$

and multiplying (5.48) by the sliding surface s_z leads to

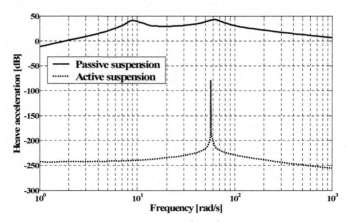

Fig. 5.8. Magnitude of $\left|\frac{\ddot{z}}{z_{r_{fl}}}\right|$ vs frequency

$$s_z \dot{s}_z = \bar{f}_{18}(x)s_z - k_z s_z^2. \tag{5.49}$$

Because the vehicle dynamics are bounded, a proper choice of the gain k_z (respectively k_θ and k_ϕ) will satisfy the necessary condition (2.58) for the existence of sliding mode.

With the modified control law (5.46), the frequency response of the system shows that the proposed SMC improves the ride comfort at low frequencies. The first three subfigures of Fig. 5.9 show a comparison between passive and active suspension. The solid line represents the magnitude of the loop transfer $\left|\frac{\ddot{z}}{z_{r_{fl}}}\right|$, $\left|\frac{\ddot{\theta}}{z_{r_{fl}}}\right|$ and $\left|\frac{\ddot{\phi}}{z_{r_{fl}}}\right|$ vs frequency without controller (*i.e.*, $k_\sigma = 0$, $\lambda_\sigma = 0$). The dotted, dashed, and dot-dashed lines represent the magnitude of the integrated controller loop transfer for a feedback gain $k_\sigma = 10$, $k_\sigma = 20$, and $k_\sigma = 30$ respectively (for a fixed value of $\lambda_\sigma = 25$ and for $\sigma \in \{z, \theta, \phi\}$). However, it can be seen that no improvements are achieved at or above the wheel frequency ω_0 ($\omega_0 \cong 57\ rad/s$).

The last subfigure of Fig. 5.9 shows the effect of λ_z on system heave acceleration $\ddot{z}$. For a fixed value of k_z ($k_z = 30$), the solid line represents the magnitude of $\left|\frac{\ddot{z}}{z_{r_{fl}}}\right|$ vs frequency for passive suspension. The dotted, dashed and dot-dashed lines represent the magnitude of $\left|\frac{\ddot{z}}{z_{r_{fl}}}\right|$ for active suspension with $\lambda_z = 1$, $\lambda_z = 5$, and $\lambda_z = 10$ respectively.

The Bode plot can be used as a criteria for selecting the proper matrices Λ and k of (5.46) to insure a good performance of the controlled system. The numerical values of the different controller gains used in simulation in addition to the desired trajectory parameters are given in Table 5.6.

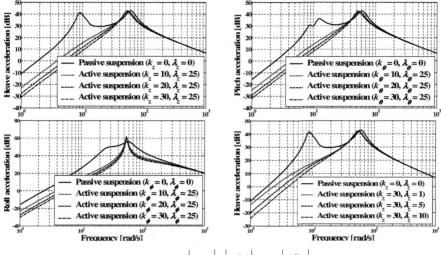

Fig. 5.9. Magnitude of $\left|\dfrac{\ddot{z}}{z_{r_{fl}}}\right|$, $\left|\dfrac{\ddot{\theta}}{z_{r_{fl}}}\right|$ and $\left|\dfrac{\ddot{\phi}}{z_{r_{fl}}}\right|$ vs frequency

Table 5.6. SMC gains and desired trajectory parameters

Parameter	Description	Value
λ_z, λ_θ, λ_ϕ	Sliding surfaces slops	25
k_z, k_θ, k_ϕ	Heave, pitch and roll gains	10
k_{2_ϑ}	Gain of module C_ϑ^c	10^4
k_{3_ϑ}	Gain of module C_ϑ^s	1
$[x_{17}^d,\ x_{19}^d,\ x_{21}^d]$	Desired trajectory parameters	[0,0,0]

5.6.2 Performance of Controlled System

The controller was tested for a road of an arbitrary form. Figure 5.10 shows the front right road input $z_{r_{fr}}$ (solid line) and the front left road input $z_{r_{fl}}$ (dashed line). The rear road inputs are similar to the front road inputs but delayed by τ_r given by $\tau_r = L/v$, where L is the distance between the front and rear axles of the vehicle ($L = a + b$) and v is the longitudinal velocity of the vehicle which is assumed to be equal to $22\ ms^{-1}$.

Performance at the Chassis

Figure 5.11 shows a comparison between passive (dashed line) and active (solid line) suspensions for heave position z and heave acceleration $\ddot{z}$.

Figure 5.12 shows a comparison between passive (dashed line) and active (solid line) suspensions for pitch angle θ and pitch angular acceleration $\ddot{\theta}$.

The roll angle ϕ and roll angular acceleration ϕ are compared for the passive (dashed line) and the active (solid line) suspensions in Fig. 5.13.

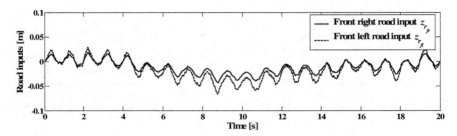

Fig. 5.10. Road inputs to the system

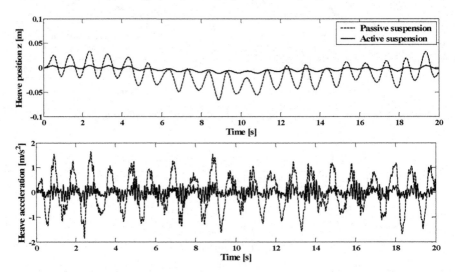

Fig. 5.11. Position z and acceleration $\ddot{z}$ for passive $(--)$ and active $(-)$ suspensions

These results are quantified using the root mean square (RMS). The RMS for a collection of n values $\{x_1, x_2, \cdots, x_n\}$ is given by

$$x_{rms} = \sqrt{\frac{1}{n}\sum_{i=1}^{n} x_i^2} = \sqrt{\frac{x_1^2 + x_2^2 + \cdots + x_n^2}{n}}. \tag{5.50}$$

Table 5.7 shows a comparison between passive and active suspensions using RMS for the chassis states.

This table also shows the percentage decrease of motions in active suspension compared to those of passive one. It is clear that active suspension improves ride comfort by reducing the effect of road perturbations on the chassis. The control is more effective for heave position and heave acceleration than for the other variables.

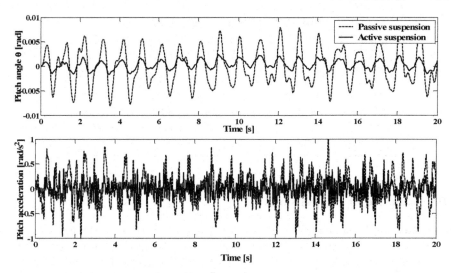

Fig. 5.12. Angle θ and acceleration $\ddot{\theta}$ for passive (– –) and active (–) suspensions

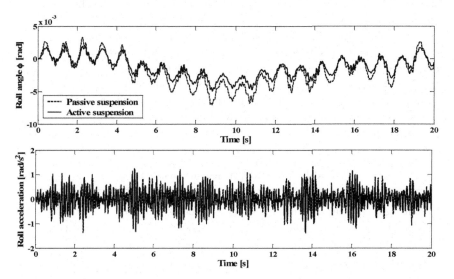

Fig. 5.13. Roll angle ϕ and roll angular acceleration $\ddot{\phi}$ for passive (– –) and active (–) suspensions

Table 5.7. RMS of the chassis states

Parameters	Passive	Active	Percentage
Heave position z	0.0257	0.0052	$- 79.76\%$
Heave acceleration $\ddot{z}$	0.7201	0.1683	$- 76.62\%$
Pitch angle θ	0.0038	0.0010	$- 73.68\%$
Pitch angular acceleration $\ddot{\theta}$	0.3484	0.1465	$- 57.95\%$
Roll angle ϕ	0.0029	0.0020	$- 31.03\%$
Roll angular acceleration $\ddot{\phi}$	0.4115	0.3858	$- 6.240\%$

Performance at Suspensions

The four suspension strokes $z_{s_\vartheta} - z_{u_\vartheta}$ ($\vartheta \in \{fr, \ fl, \ rr, \ rl\}$) for passive (dashed line) and active (solid line) suspensions are illustrated in Fig. 5.14.

It can be noted in Fig. 5.14 that the suspension strokes in passive suspension are zero-centered regardless of the road profile unlike in active suspension. For $t < 4$ s the suspension strokes are negative. This is because the controller pulls down the chassis to compensate for road convexities (see Fig. 5.10). For 6 $s < t < 18$ s, they are positive. This is because the controller pushes up the chassis to compensate for road concavities (see Fig. 5.10). This change in suspension strokes enables the system to minimize the chassis oscillation (see Figs. 5.11–5.13) which results in improving ride comfort.

Table 5.8 summarizes the RMS and the increase percentage of the suspensions stroke. The suspensions stroke are bigger for the active suspension to compensate for the road irregularities as previously explained.

Table 5.8. RMS of suspensions stroke

Variable	Passive	Active	Percentage
Front right suspension stroke $z_{s_{fr}} - z_{u_{fr}}$	0.0083	0.0159	$+91.57\%$
Front left suspension stroke $z_{s_{fl}} - z_{u_{fl}}$	0.0088	0.0188	$+113.6\%$
Rear right suspension stroke $z_{s_{rr}} - z_{u_{rr}}$	0.0071	0.0146	$+105.6\%$
Rear left suspension stroke $z_{s_{rl}} - z_{u_{rl}}$	0.0070	0.0173	$+147.1\%$

Performance at Tires

The tire deflections for passive (dashed line) and active (solid line) suspensions are illustrated in Figs. 5.15–5.18.

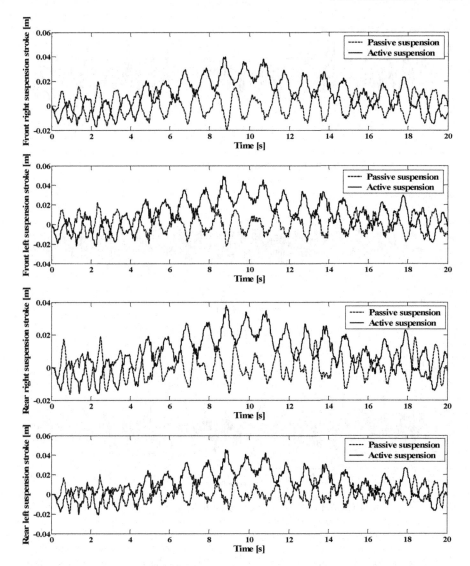

Fig. 5.14. Suspension strokes $z_{s_{fr}} - z_{u_{fr}}$, $z_{s_{fl}} - z_{u_{fl}}$, $z_{s_{rr}} - z_{u_{rr}}$, and $z_{s_{rl}} - z_{u_{rl}}$

A comparison between passive and active suspensions for the tires deflection is given in Table 5.9. The active suspension reduces the tire deflection which improves system's security and road handling.

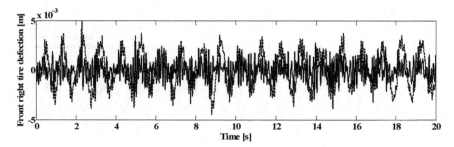

Fig. 5.15. Front right tire deflection $z_{u_{fr}} - z_{r_{fr}}$

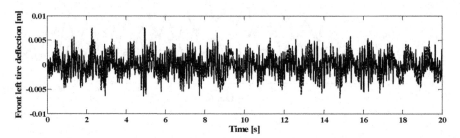

Fig. 5.16. Front left tire deflection $z_{u_{fl}} - z_{r_{fl}}$

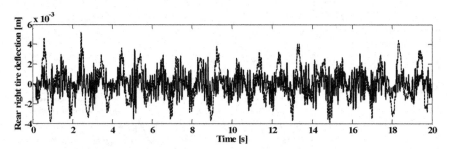

Fig. 5.17. Rear right tire deflection $z_{u_{rr}} - z_{r_{rr}}$

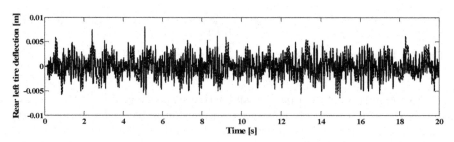

Fig. 5.18. Rear left tire deflection $z_{u_{rl}} - z_{r_{rl}}$

Table 5.9. RMS of tires deflection

Variable	Passive	Active	Percentage
Front right tire deflection $z_{u_{fr}} - z_{r_{fr}}$	0.0017	9.25×10^{-4}	$- 45.59\%$
Front left tire deflection $z_{u_{fl}} - z_{r_{fl}}$	0.0023	0.0018	$- 21.74\%$
Rear right tire deflection $z_{u_{rr}} - z_{r_{rr}}$	0.0016	9.86×10^{-4}	$- 38.37\%$
Rear left tire deflection $z_{u_{rl}} - z_{r_{rl}}$	0.0021	0.0018	$- 14.28\%$

Actuation Forces f_ϑ Generation

Figures 5.19 – 5.22 illustrate the actuation forces tracking. Simulation shows that the desired forces (dashed line) determined by the chassis control module $C_{Chassis}$ are generated (solid line) by the actuators.

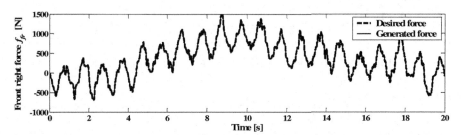

Fig. 5.19. Front right desired ($--$) and generated ($-$) forces f_{fr}

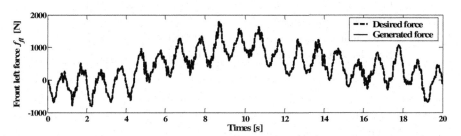

Fig. 5.20. Front left desired ($--$) and generated ($-$) forces f_{fl}

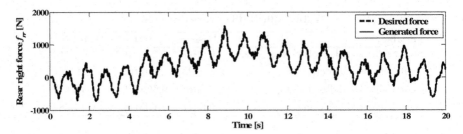

Fig. 5.21. Rear right desired (– –) and generated (–) forces f_{rr}

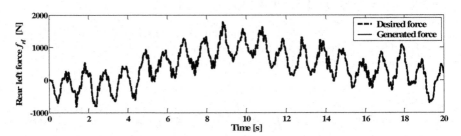

Fig. 5.22. Rear left desired (– –) and generated (–) forces f_{rl}

Table 5.10 summarizes the RMS of force tracking errors and the error percentage with respect to the desired forces.

Table 5.10. Force tracking errors

Variable	Tracking error	Percentage
Front right actuation force f_{fr}	0.4523	0.08%
Front left actuation force f_{fl}	0.8092	0.12%
Rear right actuation force f_{rr}	0.4184	0.07%
Rear left actuation force f_{rl}	0.7794	0.12%

Spool Valve Position z_{v_ϑ} Generation

Figures 5.23 – 5.26 show the spool valve position z_{v_ϑ} tracking for the active suspension. The desired positions (dashed line) determined by the control module C_ϑ^c are generated (solid line) by the actuators.

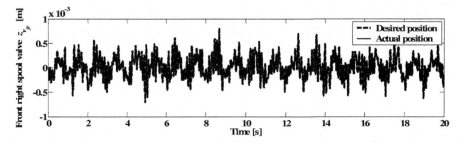

Fig. 5.23. Front right desired (– –) and generated (–) spool valve position $z_{v_{fr}}$

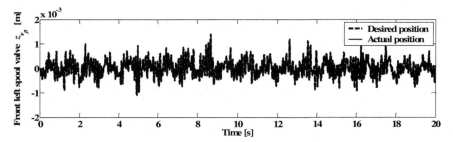

Fig. 5.24. Front left desired (– –) and generated (–) spool valve position $z_{v_{fl}}$

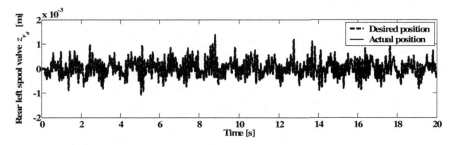

Fig. 5.25. Rear right desired (– –) and generated (–) spool valve position $z_{v_{rr}}$

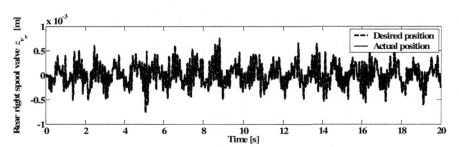

Fig. 5.26. Rear left desired (– –) and generated (–) spool valve position $z_{v_{rl}}$

Table 5.11 summarizes the tracking error of spool valve positions and the error percentage with respect to desired positions. The tracking errors are negligible.

Table 5.11. Spool valve positions tracking

Variable	Tracking error	Percentage
Front right spool valve position $z_{v_{fr}}$	1.435×10^{-6}	0.65%
Front left spool valve position $z_{v_{fl}}$	2.021×10^{-6}	0.60%
Rear right spool valve position $z_{v_{rr}}$	1.113×10^{-6}	0.51%
Rear left spool valve position $z_{v_{rl}}$	1.651×10^{-6}	0.50%

Control Inputs u_ϑ

Figures 5.27 – 5.30 show the four control inputs u_ϑ. These control inputs are generated by the control modules C_ϑ^s ($\vartheta \in \{fr,\ fl,\ rr,\ rl\}$).

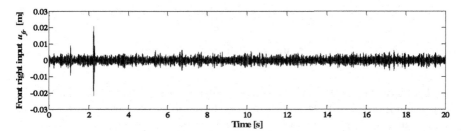

Fig. 5.27. Front right control input u_{fr}

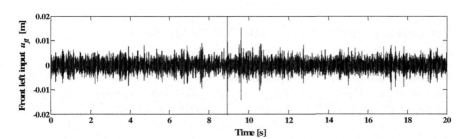

Fig. 5.28. Front left control input u_{fl}

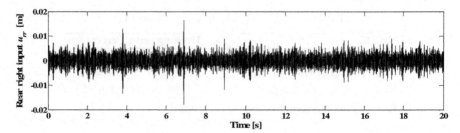

Fig. 5.29. Rear right control input u_{rr}

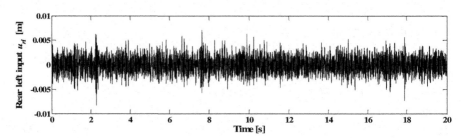

Fig. 5.30. Rear left control input u_{rl}

It should be noted that some active suspension test rigs can generate an actuation force of 9800 N [5]. Moreover, the servo valves have a time constant $\tau = 0.003$ s. Thus, the forces obtained (Figs. 5.19–5.22) and spool valve positions (Figs. 5.23–5.26) using the SMC are made possible to be generated with the existing actuators. The control inputs u_ϑ illustrated in Figs. 5.27–5.30 are given by $u_\vartheta = K_\vartheta i_\vartheta$. K_ϑ is the servo valve gain constant and i_ϑ is the input current. Being electrical inputs, u_ϑ can be generated ($\vartheta \in \{fr,\ fl,\ rr,\ rl\}$).

5.6.3 Sliding Mode Control with Model Mismatch

The controller was tested when the system and the model were mismatched. For this purpose, some model parameters were changed whereas the controller parameters were kept fixed. Table 5.12 summarizes the parameters mismatch values. The sprung mass is increased by 200 kg and the springs and tires stiffness and the suspensions damping are decreased by some percentage to illustrate an effectiveness loss or model uncertainties.

Simulation with model mismatch showed that the controller is still able to give good performance. This fact reflects the robustness of the controller which is the principal advantage of the sliding mode techniques. Tables 5.13 – 5.15 show the RMS of chassis states, suspensions stroke, and tires deflection for the passive and active suspension with and without model uncertainties. The percentage indicates the increase or the decrease of corresponding variables with respect to those of the passive suspension.

Table 5.12. System parameters mismatch

Parameter	Description	Mismatch value	Unit
M	Sprung mass	$+200$	$[kg]$
$k_{s_{fj}}$	Front springs stiffness	-5%	$[N/m]$
$k_{s_{rj}}$	Rear springs stiffness	-5%	$[N/m]$
k_{u_ϑ}	Tires stiffness	-10%	$[N/m]$
$c_{s_{fj}}$	Front suspensions damping	-2%	$[N/m/s]$
$c_{s_{rj}}$	Rear suspensions damping	-2%	$[N/m/s]$

Table 5.13. RMS of the chassis states with model uncertainties

	Without uncertainties			With uncertainties	
Variable	Passive	Active	Percentage	Active	Percentage
Heave position z	0.0257	0.0052	-79.76%	0.0057	-77.82%
Heave acceleration $\ddot{z}$	0.7201	0.1683	-76.62%	0.0959	-86.68%
Pitch angle θ	0.0038	0.0010	-73.68%	0.0010	-73.68%
Pitch angular acceleration $\ddot{\theta}$	0.3484	0.1465	-57.95%	0.0703	-79.82%
Roll angle ϕ	0.0029	0.0020	-31.03%	0.0021	-27.58%
Roll angular acceleration $\ddot{\phi}$	0.4115	0.3858	-6.240%	0.0659	-83.98%

Table 5.14. RMS of suspensions stroke with uncertainties

	Without uncertainties			With uncertainties	
Variable	Passive	Active	Percentage	Active	Percentage
Front right stroke $z_{s_{fr}} - z_{u_{fr}}$	0.0083	0.0159	$+91.57\%$	0.0153	$+84.34\%$
Front left stroke $z_{s_{fl}} - z_{u_{fl}}$	0.0088	0.0188	$+113.6\%$	0.0180	$+104.5\%$
Rear right stroke $z_{s_{rr}} - z_{u_{rr}}$	0.0071	0.0146	$+105.6\%$	0.0140	$+97.18\%$
Rear left stroke $z_{s_{rl}} - z_{u_{rl}}$	0.0070	0.0173	$+147.1\%$	0.0164	$+134.2\%$

Table 5.15. RMS of tires deflection with uncertainties

	Without uncertainties			With uncertainties	
Variable	Passive	Active	Percentage	Active	Percentage
Front right $z_{u_{fr}} - z_{r_{fr}}$	0.0017	9.25×10^{-4}	-45.59%	4.68×10^{-4}	-72.47%
Front left $z_{u_{fl}} - z_{r_{fl}}$	0.0023	0.0018	-21.74%	6.39×10^{-4}	-72.21%
Rear right $z_{u_{rr}} - z_{r_{rr}}$	0.0016	9.86×10^{-4}	-38.37%	5.38×10^{-4}	-66.37%
Rear left $z_{u_{rl}} - z_{r_{rl}}$	0.0021	0.0018	-14.28%	5.82×10^{-4}	-72.28%

Remark 5.4. In Tables 5.13 – 5.15 the performance of the control law applied to an uncertain model may appear better than those applied to a model

without uncertainties. Indeed, this is not true since some uncertainties (as for example the increase in the sprung mass) may minimize the effect of the road perturbation on the chassis.

In addition, simulation shows a small degradation in the force and spool valve position tracking. These results are illustrated in Table 5.16 for the force tracking and in Table 5.17 for the spool valve tracking.

Table 5.16. Actuation forces tracking error with model uncertainties

Variable	Without uncertainties		With uncertainties	
	Error	Percentage	Error	Percentage
Front right actuator force f_{fr}	0.4523	0.08%	0.5704	0.11%
Front left actuator force f_{fl}	0.8092	0.12%	0.7660	0.12%
Rear right actuator force f_{rr}	0.4184	0.07%	1.3043	0.25%
Rear left actuator force f_{rl}	0.7794	0.12%	0.5925	0.10%

Table 5.17. Spool valve position tracking error with model uncertainties

Variable	Without uncertainties		With uncertainties	
	Error	Percentage	Error	Percentage
Front right spool valve $z_{v_{fr}}$	1.43×10^{-6}	0.65%	3.196×10^{-6}	2.22%
Front left spool valve $z_{v_{fl}}$	2.02×10^{-6}	0.60%	4.202×10^{-6}	2.24%
Rear right spool valve $z_{v_{rr}}$	1.11×10^{-6}	0.51%	7.84×10^{-6}	5.56%
Rear left spool valve $z_{v_{rl}}$	1.65×10^{-6}	0.50%	3.194×10^{-6}	1.81%

The test of the controller in the presence of model mismatch showed that the controller is robust against uncertainties. Thus, it would give satisfactory results when applied to real systems.

5.6.4 Sensor Fault Effect

The considered fault is the breakdown of the sensor measuring the pitch angular velocity $\dot{\theta}$. This sensor is assumed to have broken down at time $t = 5\ s$. Figures 5.31 and 5.32 show, respectively, the pitch angle and the pitch angular velocity: the solid lines represent the true variables whereas the dashed lines represent the erroneous measurements. At time $t = 5\ s$, the pitch angular velocity measured by the sensor becomes zero (Fig. 5.32) and the pitch angle obtained by the integration of the measurement becomes constant (Fig. 5.31).

In the absence of a fault-tolerant system, the breakdown of this sensor mainly affects the pitch angle θ of the vehicle as shown in Fig. 5.33. When the

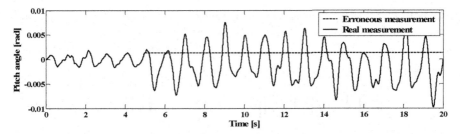

Fig. 5.31. True (–) and erroneous (..) measurement of pitch angle θ

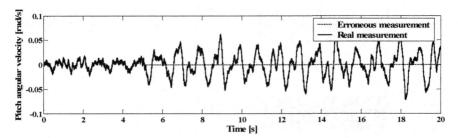

Fig. 5.32. True (–) and erroneous (..) measurement of pitch angular velocity $\dot{\theta}$

fault occurs at instant $t = 5$ s, the nominal fault-free output of the system (dashed line) is degraded (dotted line). The solid line represents the pitch angle θ for the passive suspension.

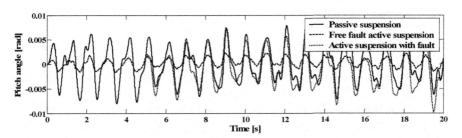

Fig. 5.33. Sensor fault effect on system's behavior

5.6.5 Fault Diagnosis

In the presence of the FDI module, the fault is detected and isolated as it has its own pattern. Figure 5.34 shows the effect of the fault on the residuals. The global FDI module compares the pattern obtained $(R_1, R_2, R_3, R_4) = (1, 1, 1, 1)$ to the *a priori* known inference matrix (Table 5.2), thus locating the fault in $SS_{chassis}$. The local FDI module of $SS_{chassis}$ (Table 5.4) examines the residuals $(R_5, R_{10}) = (1, 0)$ and isolates the faulty sensor.

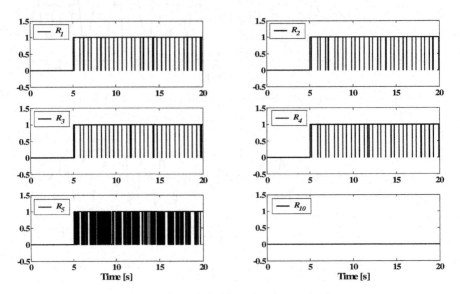

Fig. 5.34. Fault effect on the residuals

The residuals return to zero in Fig. 5.34 because the system is oscillating. This makes the value of the true oscillating variable equal or close to its faulty value during a time interval.

5.6.6 Fault-tolerance

After being detected and isolated, the FTC system module checks for the system observability and the ability to estimate the faulty measurement. In the case of the fault of this sensor, the observability is held and the estimation of the faulty measurements is possible (Table 5.5). The fault is then accommodated by injecting its estimation in the controller to replace the faulty measurement. The observer used to estimate the erroneous variables is presented in the sequel.

Estimation of Erroneous Measurements

The erroneous measurements are estimated by designing an SMO for the reduced order model $\dot{x}_{18}$, $\dot{x}_{19}$, and $\dot{x}_{20}$ given in Sect. 5.5.2. The observer dynamics are

$$\dot{\hat{x}}_{18} = \{k_{s_{fr}}x_1 + k_{s_{fl}}x_5 + k_{s_{rr}}x_9 + k_{s_{rl}}x_{13} - (k_{s_{fr}} + k_{s_{fl}} + k_{s_{rr}} + k_{s_{rl}})x_{17} - [a(k_{s_{fr}} + k_{s_{fl}}) - b(k_{s_{rr}} + k_{s_{rl}})]\sin(\hat{x}_{19}) - [d(k_{s_{fl}} + k_{s_{rl}}) - c(k_{s_{fr}} + k_{s_{rr}})]\sin(x_{21}) +$$
$$c_{s_{fr}}x_2 + c_{s_{fl}}x_6 + c_{s_{rr}}x_{10} + c_{s_{rl}}x_{14} - (c_{s_{fr}} + c_{s_{fl}} + c_{s_{rr}} + c_{s_{rl}})\hat{x}_{18} - [a(c_{s_{fr}} + c_{s_{fl}}) - b(c_{s_{rr}} + c_{s_{rl}})]\cos(\hat{x}_{19})\hat{x}_{20} - [d(c_{s_{fl}} + c_{s_{rl}}) - c(c_{s_{fr}} + c_{s_{rr}})]\cos(x_{21})x_{22} +$$
$$S(x_3 + x_7 + x_{11} + x_{15})\}/M + \lambda_{18}sign(x_{18} - \hat{x}_{18}),$$

$$\dot{\hat{x}}_{19} = \hat{x}_{20} + \lambda_{19}sign(x_{18} - \hat{x}_{18}),$$

$$\dot{\hat{x}}_{20} = \cos(\hat{x}_{19})\{ak_{s_{fr}}x_1 + ak_{s_{fl}}x_5 - bk_{s_{rr}}x_9 - bk_{s_{rl}}x_{13} - [a(k_{s_{fr}} + k_{s_{fl}}) - b(k_{s_{rr}} + k_{s_{rl}})]x_{17} - [a^2(k_{s_{fr}} + k_{s_{fl}}) + b^2(k_{s_{rr}} + k_{s_{rl}})]\sin(\hat{x}_{19}) - [d(ak_{s_{fl}} - bk_{s_{rl}}) - c(ak_{s_{fr}} - bk_{s_{rr}})]\sin(x_{21}) + ac_{s_{fr}}x_2 + ac_{s_{fl}}x_6 - bc_{s_{rr}}x_{10} - bc_{s_{rl}}x_{14} - [a(c_{s_{fr}} + c_{s_{fl}}) - b(c_{s_{rr}} + c_{s_{rl}})]\hat{x}_{18} - [a^2(c_{s_{fr}} + c_{s_{fl}}) + b^2(c_{s_{rr}} + c_{s_{rl}})]\cos(\hat{x}_{19})\hat{x}_{20} - [d(ac_{s_{fl}} - bc_{s_{rl}}) - c(ac_{s_{fr}} - bc_{s_{rr}})]\cos(x_{21})x_{22} + S[a(x_3 + x_7) - b(x_{11} + x_{15})]\}/I_{yy} + \lambda_{20}sign(x_{18} - \hat{x}_{18}).$$

When substituting the constants by their numerical values and supposing that angles are small, the estimation error dynamics become

$$\dot{\tilde{x}}_{18} = \dot{x}_{18} - \dot{\hat{x}}_{18} = -2.8\tilde{x}_{18} + 20.8\tilde{x}_{19} + 0.6267\tilde{x}_{20} - \lambda_{18}sign(\tilde{x}_{18}),$$
$$\dot{\tilde{x}}_{19} = \dot{x}_{19} - \dot{\hat{x}}_{19} = \tilde{x}_{20} - \lambda_{19}sign(\tilde{x}_{18}),$$
$$\dot{\tilde{x}}_{20} = \dot{x}_{20} - \dot{\hat{x}}_{20} = 0.4352\tilde{x}_{18} - 165.2037\tilde{x}_{19} - 4.7583\tilde{x}_{20} - \lambda_{20}sign(\tilde{x}_{18}).$$

If the gain λ_{18} is chosen such that $|20.8\tilde{x}_{19} + 0.6267\tilde{x}_{20}| \leq \lambda_{18}$:

$$\frac{d}{dt}\tilde{x}_{18}^2 = 2\tilde{x}_{18}[-2.8\tilde{x}_{18} + 20.8\tilde{x}_{19} + 0.6267\tilde{x}_{20} - \lambda_{18}sign(\tilde{x}_{18})]. \tag{5.51}$$

Then,

if $\tilde{x}_{18} > 0, \frac{d}{dt}\tilde{x}_{18}^2 = 2\tilde{x}_{18}\dot{\tilde{x}}_{18} = 2\tilde{x}_{18}(-2.8\tilde{x}_{18} + 20.8\tilde{x}_{19} + 0.6267\tilde{x}_{20} - \lambda_{18}) < 0;$

if $\tilde{x}_{18} < 0, \frac{d}{dt}\tilde{x}_{18}^2 = 2\tilde{x}_{18}\dot{\tilde{x}}_{18} = 2\tilde{x}_{18}(-2.8\tilde{x}_{18} + 20.8\tilde{x}_{19} + 0.6267\tilde{x}_{20} + \lambda_{18}) < 0.$

According to the Lyapunov principal, $\tilde{x}_{18}$ exponentially decreases to zero. If the sliding surface is defined as $s = \tilde{x}_{18} = x_{18} - \hat{x}_{18}$ then the average error dynamics during sliding when $s = 0$ and $\dot{s} = 0$ are

$$\begin{cases} \tilde{x}_{18} = 0 \\ \dot{\tilde{x}}_{18} = 0 \Rightarrow 20.8\tilde{x}_{19} + 0.6267\tilde{x}_{20} - \lambda_{18}sign(\tilde{x}_{18}) = 0 \\ \quad\quad \Rightarrow sign(\tilde{x}_{18}) = \frac{1}{\lambda_{18}}(20.8\tilde{x}_{19} + 0.6267\tilde{x}_{20}) \end{cases}, \tag{5.52}$$

and

$$\begin{cases} \dot{\tilde{x}}_{19} = \tilde{x}_{20} - \lambda_{19}sign(\tilde{x}_{18}) \\ \dot{\tilde{x}}_{20} = -165.2037\tilde{x}_{19} - 4.7583\tilde{x}_{20} - \lambda_{20}sign(\tilde{x}_{18}) \end{cases}. \tag{5.53}$$

When replacing (5.52) in (5.53), the following equations are obtained:

$$\begin{cases} \dot{\tilde{x}}_{19} = \tilde{x}_{20} - 20.8p\tilde{x}_{19} - 0.6267p\tilde{x}_{20} \\ \dot{\tilde{x}}_{20} = -165.2037\tilde{x}_{19} - 4.7583\tilde{x}_{20} - 20.8q\tilde{x}_{19} - 0.6267q\tilde{x}_{20} \end{cases}, \tag{5.54}$$

or in matrix form:

$$\begin{bmatrix} \dot{\tilde{x}}_{19} \\ \dot{\tilde{x}}_{20} \end{bmatrix} = \begin{bmatrix} -20.8p & 1 - 0.6267p \\ -165.2037 - 20.8q & -4.7583 - 0.6267q \end{bmatrix} \begin{bmatrix} \tilde{x}_{19} \\ \tilde{x}_{20} \end{bmatrix}. \tag{5.55}$$

The convergence of estimation errors $\tilde{x}_{19}$ and $\tilde{x}_{20}$ depends on the values of p and q where $p = \frac{\lambda_{19}}{\lambda_{18}}$ and $q = \frac{\lambda_{20}}{\lambda_{18}}$. These constants can be determined by pole placement such that matrix is Hurwitz. Figure 5.35 shows the estimation of the pitch angle θ and the pitch angular velocity $\dot{\theta}$. The angular velocity is estimated despite the measurement noise.

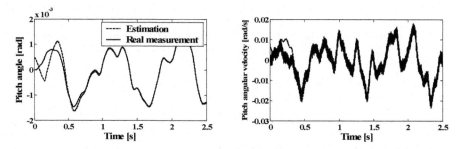

Fig. 5.35. Estimation of the erroneous measurements

Fault Compensation

The fault is compensated by replacing at time $t = 8\ s$ the erroneous measurements by their estimations. This is illustrated in Fig. 5.36. After the occurrence of the fault, and in the presence of FDI and FTC system modules, the pitch angle (dotted line) of the faulty system returns to its fault-free value (dashed line) at time $t = 10\ s$.

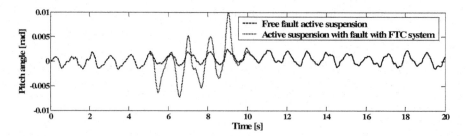

Fig. 5.36. Fault compensation

The estimations are injected in the control module at time $t = 8\ s$ by using an exponential transition law. For illustration, the fault is compensated $3\ s$ after its occurrence. The fault can be compensated faster by injecting the estimations earlier. In certain applications, the fault accommodation should be fast before the system becomes unstable.

5.7 Conclusion

This chapter presents an SMC approach for the active suspension taking into consideration the nonlinear full vehicle model and the dynamics of the four actuators. In addition to the control, an FDI module is designed to diagnose sensor faults. The system is physically broken down into subsystems whereas the control and the diagnosis structures are functionally broken down into hierarchical modules.

In order to accommodate for faults, an FTC system is designed to replace the corrupted measurement by its estimate. The idea of replacing the faulty measurement by its estimate is quite classical. However, the main idea of this chapter is in the design and the integration of the controller, the diagnosis, and the fault-tolerant modules.

Experimental application of the proposed strategy could not be made because of the absence of a test rig. However, the model of the force actuator is being validated by other researchers [2, 51].

Many of the research studies on active suspension do not take into consideration the force actuators. These works suppose that the control input desired by the controller is the control input to the system without investigating if the desired control input can be generated by the actuators or not. This assumption is not always true. Simulation shows that actuators are not always able to generate the desired forces. This usually happens for sudden or high frequency road perturbations, or for actuators saturation. The actuators consist of mechanical parts with certain response time which bounds their capacity. This point will be studied to determine, for given specifications, the operating range of the actuators.

The FDI strategy is able to detect and isolate abrupt sensor faults such as gains, bias, breakdown, or freeze. However, the isolation of sensor faults is not always precise. This possibility depends on several factors: the amplitude of the fault, the gains of the observers, the chosen thresholds, and the amplitude of the road perturbation exciting the system.

The gains of the observers should be chosen to allow fast convergence. However, a high gain may make the estimation, used in the generation of the residuals, insensitive to the fault and consequently the FDI misses its occurrence.

As previously defined, the residual is the comparison between the measured and the estimated variable. Ideally the residual equals zero when there is no fault and the deviation of the residual from its nominal value indicates the presence of a fault. However, due to measurement noise and model uncertainties, the residual does not equal zero in fault-free mode. Thus adequate thresholds should be defined to prevent fault alarms of the presence of a fault and at the same time to prevent the FDI missing the occurrence of faults. Many researchers treated the problem of designing robust and dynamic thresholds for the purpose of FDI for uncertain systems [75].

The decision about the sensor fault location depends on its signature. However, a sufficient time interval should be allowed after the detection of the fault to permit a correct isolation.

6

Conclusion

The theory of FDI and FTC has been developed for years. The objective of this book was to present FDI/FTC techniques applied to real laboratory-scale systems or in simulation to a realistic model of an industrial system.

Many research studies have been developed in the literature. Only some of them are applied to real systems and many others consider theoretical development in linear and nonlinear cases. Moreover, many of these methods do not take into account the fault diagnostic module and consider that the fault has already been detected and isolated properly. However, it goes without saying that the performance of any FTC method is tightly linked to the information issued from the FDI module.

The methods presented in this book are intended as a contribution to the application of FDI/FTC methods to real systems. One of the advantages of these approaches is that they consider the whole steps of an FTC method. A guideline is given to develop a complete method emphasizing the importance of the selection of an operating point and including modeling, identification, nominal tracking control, fault diagnosis and the FTC design.

Chapter 2 presented a theoretical development of an FDI/FTC approach in the linear case and then in the nonlinear case. Through many years of teaching and developing research activities, it was noticed that the notion of the operating point is still ambiguous and not easily understood. One objective of this book was to clarify this issue which is very important in the design of a nominal control when the system has to be linearized around an operating point. The modeling of sensor and actuator faults has been recalled and the analysis of their effect on the system has been detailed. Then, the nominal tracking control method is reexamined. The performance of the FTC method presented later in the chapter has to be as close as possible to the performance of the nominal tracking control. The FDI is a major issue in FTC design. That is why attention has been paid to the presentation of an FDI module which is integrated to the whole strategy. Once the fault is detected and isolated, two different approaches are presented to estimate the fault magnitude. This estimation is then used to compensate for the fault effect.

In parallel to the linear study, an FDI/FTC method based on nonlinear techniques has been detailed. This nonlinear approach takes into account all required steps for fault tolerance design. Finally, the occurrence of major failures such as the complete loss of a sensor or an actuator was discussed and a method of dealing with the presence of such failures was proposed. If a sensor is completely lost, the compensation method using the additive control law is still able to compensate for the sensor fault if the system is still observable despite the sensor loss.

Regarding the blocking or the complete loss of an actuator, if there is no hardware redundancy, the system becomes uncontrollable. In this case, the system is restructured in order to distribute the control law over the healthy actuators.

Chapter 3 is dedicated to apply the complete FDI/FTC approaches described in Chap. 2 to a winding machine which is a nonlinear system. This system was first considered as linear around an operating point. Then, the nominal tracking control, the FDI, and the FTC for sensor and actuator faults are designed and detailed. This application gives the reader a detailed example taking into account all the steps to follow in order to design an FDI/FTC method if the system is described by a linear state-space representation.

As stated previously, the model of the winding machine is nonlinear. However, the description of the nonlinear model is difficult to achieve. Therefore, a modeling of the system for several operating zones was proposed. A detailed FTC method was applied to the winding machine throughout the whole operating zone.

In Chap. 4, a three-tank system was considered to illustrate the FDI/FTC approaches in both linear and nonlinear cases. A detailed study was conducted and discussed in the presence of sensor and actuator faults. While for the winding machine the linearized model was obtained experimentally, the nonlinear model of the three-tank system is easy to get and to linearize around an operating point using, for instance, the Taylor expansion. The methods developed in Chap. 2 were applied to this system in the linear case for faults such as biases or drifts on the measurements or a loss of actuators effectiveness.

The case of major failures was discussed as well. The objective was to keep the system operating safely by redefining the degraded performance to reach if one actuator is blocked or completely lost. A new equilibrium point was calculated based on the nonlinear equations in order to avoid system shutdown. Similar study for the major actuator failures with application to an unmanned aerial vehicle has been conducted and showed interesting results in calculating a new trim point if a control surface of the aircraft is blocked. For more details, the reader can refer to [12].

A full vehicle active suspension system was used in simulation to present a detailed FDI/FTC approach in Chap. 5. The particularity of this system is that it is a complex system. It is described by a nonlinear model of order 22. Since it is difficult to deal with this high order model to design a nominal controller, an FDI module, and an FTC method, the system is broken down into five interconnected subsystems. Each subsystem is driven by a local controller and monitored by a local FDI module. A higher level coordinates the information issued from these local modules. For this system, only sensor faults were considered. The nominal control and the state estimation were based on the sliding mode techniques. The objective was to substitute the faulty or the lost measurement by its estimation. This idea is not new; however the challenge here was in the way of estimating the state variables and the outputs for this high order system.

Another advantage of this study in Chap. 5 corresponded to the technological analysis of the available sensors on the market. This is an interesting issue in real applications where one has to optimize the number of sensors to use from the economical point of view and the available space to install these sensors. Once a sensor is lost, the objective is to estimate the corresponding output using available and healthy measurement. This requires that the system is still observable despite the sensor loss. A study has been conducted to determine the minimal number of sensors to use in a complex system while ensuring its observability. This study was not only developed in the nominal case, but it also aimed at determining the optimal number of sensors to install in order to keep the system observable despite the loss of a sensor. For more details, the reader can refer to [18].

This book does not aim at being an exhaustive overview of FDI/FTC methods. For instance, one way to consider a wide operating zone could consist of using multiple-model techniques rather than considering an exact nonlinear model. The reader can refer to [106] where this modeling technique is used to provide a specific FDI approach applied to the three-tank system. As recently considered in the literature, FDI/FTC methods have been developed under linear parameter-varying (LPV) model representation in order to describe the dynamic behavior of systems. In this way, as recently proposed by [123], the winding machine can also be considered as an LPV system with an appropriate dedicated FDI/FTC strategy. Moreover, the objective of the book was to show some applications developed for many years. Among these applications, unmanned aerial vehicles, not presented in this book, is an interesting system to show a detailed FDI/FTC approach with challenging problems in the presence of major actuator failures such as the blocking of a control surface [11]. Finally, the robustness of a given solution against FDI inaccuracies which is a key problem in FTC methods [141], should be addressed. thods [141], should be addressed.

A

Three-tank System Simulation

The three-tank system is simulated using MATLAB®/Simulink®. The simulation has been developed using MATLAB® version 6.5.1 (R13SP1), Simulink® version 5.1 (R13SP1), and Virtual Reality Toolbox® version 3.1 (R13SP1). It consists of the modeling part and the animation part. It also works with newer MATLAB® versions such as version 7.3 (R2006b).

A.1 Main Page

The main Simulink® file is *Three_tank.mdl*. Once this file is open, the main page illustrated in Fig. A.1 appears. The reference levels are displayed on the right hand side of this page in addition to the sensors measurement and the real levels in the tank.

It is also possible to test the effect of sensor and actuator faults on the behavior of the system by using the blocks on the left hand side of the main page. Consider for example the block shown in Fig. A.2. This block allows us to simulate bias faults for sensor 1 by adjusting the fault amplitude (here it is set to −0.03) and the fault time occurrence (set to 1500 s). Users can easily test other faults such as drifts or freezing.

Figure A.3 shows the block that allows us to simulate a fault on actuator 1. In this case, the *Loss* block is set to 0.2 which means that a loss of effectiveness of 20% is supposed to occur at 1500 s. A value of 1 means the complete loss of the actuator. A value of 0 means that no fault occurs.

Finally, the main page contains the animation window which shows up automatically once simulation is started. This window will be explained in Sect. A.3. It should be noted that the model is automatically initialized; thus a manual initialization is not needed.

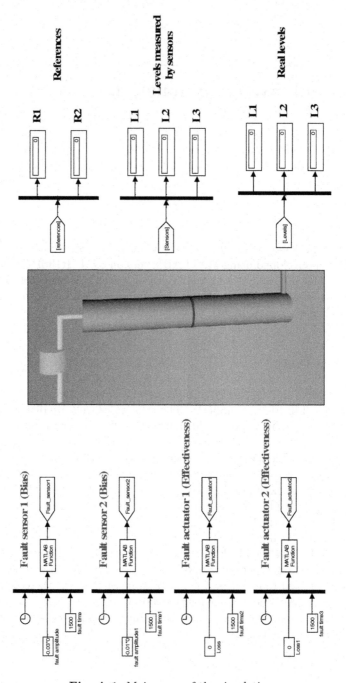

Fig. A.1. Main page of the simulation

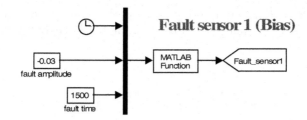

Fig. A.2. Sensor fault block

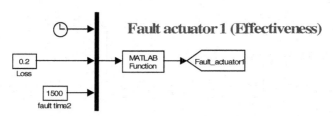

Fig. A.3. Actuator fault block

A.2 Modeling Part

The modeling part shown in Fig. A.4 consists of:

- The nonlinear model of the system (the block *"Three tank system"*)
- The controller
- The reference levels block
- Two blocks to simulate sensor and actuator faults
- Signal routing blocks for the animation purposes

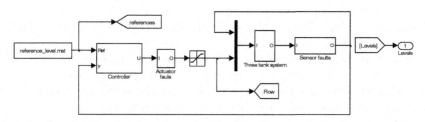

Fig. A.4. The three-tank system

The control law used in this simulation is a state-feedback with integrator where the gains are determined using a linearized model around an operating point (Fig. A.5). Therefore, the controller will give satisfactory performance only around this operating point. Thus, users should pay attention not to drive the system outside the operating region when simulating the nominal behavior of the system. On the other hand, users are invited to apply nonlinear control laws (see for example Chap. 2) for the whole operating range of the system.

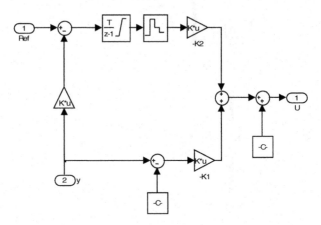

Fig. A.5. State-feedback with integrator controller

Remark A.1. It should be noted that, in this simulation, actuators are assumed to be scalar gains. In addition, no sensor noise is used. These two points can be easily considered in simulation for more consistency.

A.3 Animation Part

The animation window shown in Fig. A.6 allows one to visualize the measurements issued from the modeling part. As stated before, this window shows up automatically when simulation starts, but it can also be forced to show up by clicking the middle block of the main page (Fig. A.1).

This window shows the three interconnected tanks, the two pumps, and two red rings representing the reference levels (set-points). During the simulation, users can note how water levels follow the references and how the different water flow rates vary with time. This animation is not only useful in displaying the measurements, but also in examining what is going on in the real system when faults occur. In the case of a sensor fault (a bias for example), the sensor tells that the reference is followed while this is not the case in reality. This can be easily seen on the animation.

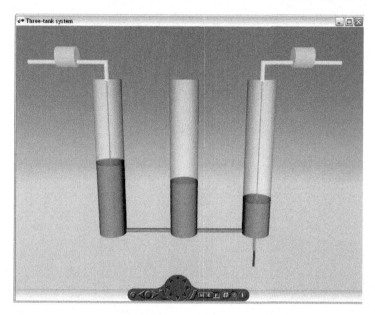

Fig. A.6. The animation window

A.4 Various Files

This simulation comes with a set of files:

- *Three_tank.mdl*: the main simulation file
- *init_para.m*: file containing the different constants initializing the model
- *tank_system.m*: the nonlinear model of the system
- *sensor_fault.m*: used to simulate sensor faults
- *actuator_fault.m*: used to simulate actuator faults
- *reference_level.mat*: a .mat file containing the reference levels
- *3tanks.wrl*: the virtual reality file used for the animation
- *tanks.bmp*: a .bmp figure used for the main page

References

1. A. Akhenak, M. Chadli, D. Maquin, and J. Ragot. State estimation via multiple observer the three-tank system. In *5th IFAC Symposium on Fault Detection, Supervision and Safety for Technical Processes*, pages 1227–1232, Washington DC, USA, 2003.
2. A. Alleyne and J.K. Hedrick. Nonlinear adaptive control of active suspensions. *IEEE Transactions on Control Systems Technology*, 3(1):94–101, 1995.
3. A. Alleyne, P.D. Neuhaus, and J.K. Hedrick. Application of nonlinear control theory to electronically controlled suspensions. *Vehicle System Dynamics*, 22(5-6):309–320, 1993.
4. Bismarckstr Amira GmbH. Documentation of the three-tank system. *65, D-47057 Duisburg, Germany*, 1994.
5. Y. Ando and M. Suzuki. Control of active suspension systems using the singular perturbation method. *Control Engineering Practice*, 4(3):287–293, 1996.
6. P.J. Antsaklis and A.N. Michel. *A linear systems primer*. A Birkhäuser Boston, 2007.
7. L. El Bahir and M. Kinnaert. Fault detection and isolation for a three-tank system based on a bilinear model of the supervised process. In *United Kingdom Automatic Control Council International Conference on Control*, volume 2, pages 1486–1491, Swansea, UK, 1998.
8. M. Basseville and I. Nikiforov. *Detection of abrupt changes: Theory and application*. Information and System Sciences Series, Prentice Hall International, 1993.
9. A. Bassong-Onana, M. Darouach, and G. Krzakala. Optimal estimation of state and inputs for stochastic dynamical systems with unknown inputs. In *Proceedings of International Conference on Fault Diagnosis*, pages 267–275, Toulouse, France, 1993.
10. T. Bastogne, H. Noura, P. Sibille, and A. Richard. Multivariable identification of a winding process by subspace methods for tension control. *Control Engineering Practice*, 6(9):1077–1088, 1998.
11. F. Bateman, H. Noura, and M. Ouladsine. Actuators fault diagnosis and tolerant control for an unmanned aerial vehicle. In *IEEE Multi-conference on Systems and Control*, pages 1061–1066, Singapore, 2007.
12. F. Bateman, H. Noura, and M. Ouladsine. A fault tolerant control strategy for a UAV based on a Sequential Quadratic Programming algorithm. In *Proceedings*

of the IEEE Conference on Decision and Control, pages 423–428, Cancun, Mexico, 2008.

13. H. Benitez-Perez and F. Garcia-Nocetti. *Reconfigurable distributed control*. London, UK: Springer, 2005.

14. M. Blanke, M. Kinnaert, J. Lunze, and M. Staroswiecki. *Diagnosis and fault-tolerant control*. Berlin, Germany: Springer, second edition, 2006.

15. M. Börner, H. Straky, T. Weispfenning, and R. Isermann. Model based fault detection of vehicle suspension and hydraulic brake systems. *Mechatronics*, 12(8):999 – 1010, 2002.

16. P. Borodani, L. Gortan, R. Librino, and M. Osella. Minimum risk evaluation methodology for fault tolerant automotive control systems. In *Proceedings of the International Conference on Applications of Advanced Technologies in Transportation Engineering*, pages 552–557, Capri, Italy, 1995.

17. F. Caccavale and L. Villani. *Fault diagnosis and fault tolerance for mechatronic systems: Recent advances*, volume 1 of *Springer tracts in advanced robotics*. Berlin, Germany: Springer, 2003.

18. A. Chamseddine, H. Noura, and M. Ouladsine. Design of minimal and tolerant sensor networks for observability of vehicle active suspension. *To appear in IEEE Transactions on Control Systems Technology*, 2009.

19. A. Chamseddine, H. Noura, and T. Raharijaona. Optimal sensor network design for observability of active suspension. *International Modeling and Simulation Multiconference*, February 2007. Buenos Aires, Argentina.

20. A. Chamseddine, H. Noura, and T. Raharijaona. Optimal sensor network design for observability of complex systems. In *Proceedings of the IEEE American Control Conference*, pages 1705–1710, New York City, USA, July 2007.

21. S. Chantranuwathana and H. Peng. Adaptive robust force control for vehicle active suspensions. *International Journal of Adaptive Control and Signal Processing*, 18(2):83–102, 2004.

22. J. Chen and R.J. Patton. *Robust model-based fault diagnosis for dynamic systems*. Kluwer academic publishers, 1999.

23. P.C. Chen and A.C. Huang. Adaptive sliding control of non-autonomous active suspension systems with time-varying loadings. *Journal of Sound and Vibration*, 282(3-5):1119–1135, 2005.

24. L. Chiang, E. Russell, and R. Braatz. *Fault detection and diagnosis in industrial systems*. New-York, USA: Springer-Verlag, 2001.

25. T.-H. Chien, J. S.-H. Tsai, S.-M. Guo, and J.-S. Li. Low-order self-tuner for fault-tolerant control of a class of unknown nonlinear stochastic sampled-data systems. *Applied Mathematical Modelling*, 33(2):706–723, 2008.

26. K.C. Chiou and S.J. Huang. An adaptive fuzzy controller for 1/2 vehicle active suspension systems. In *IEEE International Conference on Systems, Man and Cybernetics*, volume 2, pages 1010–1015, Taipei, Taiwan, 2006.

27. F.J. D'Amato and D.E. Viassolo. Fuzzy control for active suspensions. *Mechatronics*, 10(8):897–920, 2000.

28. M. Darouach, M. Zasadzinski, and M. Hayar. Reduced-order observer design for descriptor systems with unknown inputs. *IEEE Transactions on Automatic Control*, 41(7):1068–1072, 1996.

29. J. D'Azzo and C.H. Houpis. *Linear control system analysis and design, conventional and modern*. McGraw-Hill Series in Electrical and Computer Engineering, 1995.

30. P. Van den Hof. Closed-loop issues in system identification. *Annual Reviews in Control*, 22:173–186, 1998.

31. S.X. Ding. *Model-based fault diagnosis techniques - design schemes, algorithms and tools*. Springer, 2008.

32. H. Du, J. Lam, and K.Y. Sze. Non-fragile output feedback H_∞ vehicle suspension control using genetic algorithm. *Engineering Applications of Artificial Intelligence*, 16(7):667–680, 2003.

33. H. Du and N. Zhang. H_∞ control of active vehicle suspensions with actuator time delay. *Journal of Sound and Vibration*, 301(1):236–252, 2007.

34. D. Fischer, M. Börner, J. Schmitt, and R. Isermann. Fault detection for lateral and vertical vehicle dynamics. *Control Engineering Practice*, 15(3):315–324, 2007.

35. D. Fischer and R. Isermann. Mechatronic semi-active and active vehicle suspensions. *Control Engineering Practice*, 12(11):1353–1367, 2004.

36. D. Fischer and R. Isermann. Model based process fault detection for a vehicle suspension actuator. In *7th International Symposium on Advanced Vehicle Control*, pages 573–578, Barcelona, Spain, August 2004.

37. D. Fischer, E. Kaus, and R. Isermann. Fault detection for an active vehicle suspension. In *Proceedings of the IEEE American Control Conference*, volume 5, pages 4377–4382, Denver, Colorado, USA, June 2003.

38. D. Fischer, E. Kaus, and R. Isermann. Model based sensor fault detection for an active vehicle suspension. In *5th IFAC Symposium on Fault Detection, Supervision and Safety of Technical Processes*, Washington DC, USA, June 2003.

39. D. Fischer, H.P. Schöner, and R. Isermann. Model based fault detection for an active vehicle suspension. In *World Automotive Congress*, Barcelona, Spain, May 2004.

40. D. Fischer, M. Zimmer, and R. Isermann. Identification and fault detection of an active vehicle suspension. In *13th IFAC Symposium on System Identification*, Rotterdam, Netherland, August 2003.

41. A.J. Fossard and D. Normand Cyrot. *Nonlinear systems*, volume 1 of *Modeling and estimation*. Chapman & Hall, London, New York, 1995.

42. A.L. Fradkov, I.V. Miroshnik, and V.O. Nikiforov. *Nonlinear and adaptive control of complex systems*. Kluwer Academic Publishers, 1999.

43. P.M. Frank. Fault diagnosis in dynamic systems using analytical and knowledge based redundancy - a survey of some new results. *Automatica*, 26(3):459–474, 1990.

44. P.M. Frank, S.X. Ding, and B. Kppen-Seliger. Current developments in the theory of FDI. In *4th IFAC Symposium on Fault Detection, Supervision and Safety of Technical Processes*, pages 16–27, Budapest, Hungary, 2000.

45. G.F. Franklin, J.D. Powell, and M.L. Workman. *Digital Control of Dynamic Systems*. Addison-Wesley Ed., second edition, 1990.

46. R.A. Freeman and P.V. Kokotovic. *Robust nonlinear control design: State-space and Lyapunov techniques*. Birkhäuser Boston, 2008.

47. Z. Gao and P.J. Antsaklis. Stability of the pseudo-inverse method for reconfigurable control systems. *International Journal of Control*, 53(3):717–729, 1991.

48. Z. Gao and P.J. Antsaklis. Reconfigurable control system design via perfect model following. *International Journal of Control*, 56(4):783–798, 1992.

49. E. Alcorta Garcia and P.M. Frank. Deterministic nonlinear observer based approaches to fault diagnosis: a survey. *Control Engineering Practice*, 5(5):663–670, 1997.

50. E. Alcorta Garcia and P.M. Frank. Fault detection and isolation in nonlinear systems. In *Proceedings of the European Control Conference*, Karlsruhe, Germany, 1999. CD-Rom.

51. P. Gáspár, I. Szászi, and J. Bokor. Mixed H_2/H_∞ control design for active suspension structures. *Periodica Polytechnica - Transportation Engineering*, 28(1-2):3–16, 2000.

52. O. Gasparyan. *Linear and nonlinear multivariable feedback control: A classical approach*. Wiley, 2008.

53. J.J. Gertler. *Fault detection and diagnosis in engineering systems*. Marcel Dekker, Inc. New York Basel Hong Kong, 1998.

54. G.H. Golub and C.F. Van Loan. *Matrix computations*. The Johns Hopkins University Press, second edition, 1989.

55. G.C. Goodwin and K.S. Sin. *Adaptive filtering prediction and control*. Prentice-Hall information & system sciences series, 1984.

56. R. Güçlü. Active control of seat vibrations of a vehicle model using various suspension alternatives. *Turkish Journal of Engineering and Environmental Sciences*, 27:361–373, 2003.

57. R. Güçlü and K. Gulez. Neural network control of seat vibrations of a nonlinear full vehicle model using PMSM. *Mathematical and Computer Modelling*, 47(11-12):1356–1371, 2008.

58. W.M. Haddad and V. Chellaboina. *Nonlinear dynamical systems and control: A Lyapunov-based approach*. Princeton University Press, 2008.

59. C. Hajiyev and F. Caliskan. *Fault diagnosis and reconfiguration in flight control systems*. London, UK: Kluwer Academic Publishers, 2003.

60. H. Hammouri, P. Kabore, and M. Kinnaert. A geometric approach to fault detection and isolation for bilinear systems. *IEEE Transactions on Automatic Control*, 46(9):1451–1455, 2001.

61. T.J. Harris, C. Seppala, and L.D. Desborough. A review of performance monitoring and assessment techniques for univariate and multivariate control systems. *Journal Process of Control*, 9(1):1–17, 1999.

62. B. Heiming and J. Lunze. Definition of the three-tank benchmark problem for controller reconfiguration. In *Proceedings of the European Control Conference*, Karlsruhe, Germany, 1999. CD-Rom.

63. D. Henry and A. Zolghadri. Norm-based design of robust FDI schemes for uncertain systems under feedback control: Comparison of two approaches. *Control Engineering Practice*, 14(9):1081–1097, 2006.

64. D. Henry, A. Zolghadri, M. Monsion, and S. Ygorra. Off-line robust fault diagnosis using the generalized structured singular value. *Automatica*, 38(8):1347–1358, 2002.

65. D. Hinrichsen and D. Prätzel-Wolters. *A canonical form for static linear output feedback*. Lecture Notes in Control and Information Sciences. Springer Berlin/Heidelberg, 1984.

66. M. Hou and P.C. Muller. Disturbance decoupled observer design: a unified viewpoint. *IEEE Transactions on Automatic Control*, 39(6):1338–1341, 1994.

67. S. Ikenaga, F.L. Lewis, J. Campos, and L. Davis. Active suspension control of ground vehicle based on a full-vehicle model. In *Proceedings of the IEEE American Control Conference*, pages 4019–4024, Chicago, Illinois, USA, 2000.

68. H. Imine, N.K. M'Sirdi, and Y. Delanne. Adaptive observers and estimation of the road profile. *SAE Transactions - Journal of Passenger Cars - Mechanical Systems*, 112(6):1312–1317, 2003.

69. R. Isermann. *Fault-Diagnosis Systems: an introduction from fault detection to fault tolerance.* Springer, 2006.

70. R. Isermann and P. Ballé. Trends in the application of model-based fault detection and diagnosis of technical processes. In *13th IFAC World Congress*, pages 1–12, San Francisco, USA, 1996.

71. R. Isermann and P. Ballé. Trends in the application of model-based fault detection and diagnosis of technical processes. *Control Engineering Practice*, 5(5):709–719, 1997.

72. R. Isermann, D. Fischer, M. Börner, and J. Schmitt. Fault detection for lateral and vertical vehicle dynamics. In *3rd IFAC Symposium on Mechatronic Systems*, Sidney, Australia, September 2004.

73. A. Isidori. *Nonlinear control systems.* Springer-Verlag, third edition, 1995.

74. B.P. Jeppesen and D. Cebon. Analytical redundancy techniques for fault detection in an active heavy vehicle suspension. *Vehicle System Dynamics*, 42(1-2):75–88, 2004.

75. A. Johansson and M. Bask. Dynamic threshold generators for fault detection in uncertain systems. In *16th IFAC World Congress*, Prague, Czech Republic, July 2005.

76. C. Join, J.-C. Ponsart, D. Sauter, and D. Theilliol. Nonlinear filter design for fault diagnosis: application to the three-tank system. *IEE Proceedings - Control Theory and Applications*, 152(1):55–64, 2005.

77. K. Kashi, D. Nissing, D. Kesselgruber, and D. Soffker. Fault diagnosis of an active suspension control system. In *6th IFAC Symposium on Fault Detection, Supervision and Safety of Technical Processes*, pages 535–540, Beijing, P.R. China, 2006.

78. T. Kaylath. *Linear systems.* Prentice-Hall, 1980.

79. J.Y. Keller. Fault isolation filter design for linear stochastic systems. *Automatica*, 35(10):1701–1706, 1999.

80. M. Kinnaert. Robust fault detection based on observers for bilinear systems. *Automatica*, 35(11):1829–1842, 1999.

81. M. Kinnaert. Fault diagnosis based on analytical models for linear and nonlinear systems: a tutorial. In *5th IFAC Symposium on Fault Detection, Supervision and Safety for Technical Processes*, pages 37–50, Washington DC, USA, 2003.

82. D. Koenig, S. Nowakowski, and T. Cecchin. An original approach for actuator and component fault detection and isolation. In *3rd IFAC Symposium on Fault Detection Supervision and Safety for Technical Processes*, pages 95–105, Hull, UK, 1997.

83. J.M. Koscielny. Application of fuzzy logic for fault isolation in a three-tank system. In *14th IFAC World Congress*, pages 73–78, Beijing, R.P. China, 1999.

84. I. E. Köse and F. Jabbari. Scheduled controllers for linear systems with bounded actuators. *Automatica*, 39(8):1377–1387, 2003.

85. C. Lauwerys, J. Swevers, and P. Sas. Robust linear control of an active suspension on a quarter car test-rig. *Control Engineering Practice*, 13(5):577–586, 2005.

86. S. Leonhardt and M. Ayoubi. Methods of fault diagnosis. *Control Engineering and Practice*, 5(5):683–692, 1997.

87. L. Li and D. Zhou. Fast and robust fault diagnosis for a class of nonlinear systems: detectability analysis. *Computers and Chemical Engineering*, 28(12):2635–2646, 2004.

88. B. Litkouhi and N.M. Boustany. On-board sensor failure detection of an active suspension system using the generalized likelihood ratio approach. In *Proceedings of the IEEE Conference on Decision and Control Including The Symposium on Adaptive Processes*, pages 2358–2363, Austin TX, USA, 1988.

89. Z. Liu, C. Luo, and D. Hu. Active suspension control design using a combination of LQR and backstepping. In *Chinese Control Conference*, pages 123–125, Harbin, China, 2006.

90. L. Ljung. *System identification: Theory for the user.* Prentice-Hall Inc, Englewood Cliffs, NJ, 1987.

91. C.J. Lopez and R.J. Patton. Takagi-sugeno fuzzy fault-tolerant control for a non-linear system. In *Proceedings of the IEEE Conference on Decision and Control*, pages 4368–4373, Phoenix, Arizona, USA, 1999.

92. M. Mahmoud, J. Jiang, and Y. Zhang. *Active fault tolerant control systems: Stochastic analysis and synthesis (Lecture notes in control and information sciences).* Berlin, Germany: Springer, 2003.

93. T. Marcu, M.H. Matcovschi, and P.M. Frank. Neural observer-based approach to fault-tolerant control of a three-tank system. In *Proceedings of the European Control Conference*, Karlsruhe, Germany, 1999. CD-Rom.

94. S. Marzbanrad, G. Ahmadi, H. Zohoor, and Y. Hojjat. Stochastic optimal preview control of a vehicle suspension. *Journal of Sound and Vibration*, 275(3-5):973–990, 2004.

95. L.F. Mendonca, J. Sousa, and J.M.G. Sa da Costa. Fault accommodation of an experimental three-tank system using fuzzy predictive control. In *IEEE International Conference on Fuzzy Systems*, pages 1619–1625, Hong Kong, 2008.

96. P. Metallidis, G. Verros, S. Natsiavas, and C. Papadimitriou. Fault detection and optimal sensor location in vehicle suspensions. *Journal of Vibration and Control*, 9(3-4):337–359, 2003.

97. E.A. Misawa and J.K. Hedrick. Nonlinear observer a state of the art: survey. *Journal of Dynamic Systems, Measurement, and Control Transactions*, 111(3):344–352, 1989.

98. H. Nijmeier and A.J. Van der Schaft. *Nonlinear dynamical control systems.* Springer, third edition, 1996.

99. H. Noura, D. Sauter, F. Hamelin, and D. Theilliol. Fault-tolerant control in dynamic systems: application to a winding mahine. *IEEE Control Systems Magazine*, 20(1):33–49, 2000.

100. J. Park, G. Rizzoni, and W.B. Ribbens. On the representation of sensor faults in fault detection filters. *Automatica*, 30(11):1793–1795, 1994.

101. C. De Persis and A. Isidori. A geometric approach to nonlinear fault detection. *IEEE Transactions on Automatic Control*, 46(6):853–865, 2001.

102. J.-C. Ponsart, D. Sauter, and D. Theilliol. Control and fault diagnosis of a winding machine based on a ltv model. In *Proceedings of the IEEE Conference on Control Application*, Toronto, Canada, 2005. CD-Rom.

103. J.-C. Ponsart, D. Theilliol, and H. Noura. Fault-tolerant control of a nonlinear system. application to a three-tank system. In *Proceedings of the European Control Conference*, Karlsruhe, Germany, 1999. CD-Rom.

104. C. Rago, R. Prasanth, R.K. Mehra, and R. Fortenbaugh. Failure detection and identification and fault tolerant control using the IMMF-KF with applications to eagle-eye UAV. In *Proceedings of the IEEE Conference on Decision and Control*, pages 4208–4213, Tampa, Florida, USA, 1998.

105. J. Juan Rincon-Pasaye, R. Martinez-Guerra, and A. Soria Lopez. Fault diagnosis in nonlinear systems: an application to a three-tank system. In *Proceedings of the IEEE American Control Conference*, pages 2136–2141, Washington DC, USA, 2008.

106. M. Rodrigues, D. Theilliol, M.A. Medina, and D. Sauter. A fault detection and isolation scheme for industrial systems based on multiple operating models. *Control Engineering Practice*, 16(2):225–239, 2008.

107. M.H. Sadeghi and S.D. Fassois. On-board fault identification in an automobile fully-active suspension system. In *Advances in automotive control: a postprint volume from the IFAC Workshop*, pages 139–144, Ascona, Switzerland, 1995.

108. M.A. Sainz, J. Armengol, and J. Vehi. Fault detection and isolation of the three-tank system using the modal interval analysis. *Journal of Process Control*, 12(2):325–338, 2002.

109. Y. Md. Sam, J.H.S. Osman, M. Ruddin, and A. Ghani. A class of proportional-integral sliding mode control with application to active suspension system. *Systems and Control Letters*, 51(3-4):217–223, 2004.

110. Y.M. Sam, M.R.H.A. Ghani, and N. Ahmad. LQR controller for active car suspension. In *TENCON Proceedings*, pages 441–444, Kuala Lumpur, Malaysia, 2000.

111. D. Sauter, H. Jamouli, J-Y. Keller, and J.-C. Ponsart. Actuator fault compensation for a winding machine. *Control Engineering Practice*, 13(10):1307–1314, 2005.

112. D.N. Shields and S. Du. An assessment of fault detection methods for a benchmark system. In *4th IFAC Symposium on Fault Detection Supervision and Safety for Technical Processes*, pages 937–942, Budapest, Hungary, 2000.

113. E. Silani, D. Fischer, S.M. Savaresi, E. Kaus, R. Isermann, and S. Bittanti. Fault tolerant filtering in active vehicle suspensions. In *World Automotive Congress*, Barcelona, Spain, May 2004.

114. G.K. Singh and K.E. Holé. Guaranteed performance in reaching mode of sliding mode controlled systems. *Sadhana*, 29(1):129–141, February 2004.

115. E. N. Skoundrianos and S. G. Tzafestas. Fault diagnosis via local neural networks. *Mathematics and Computers in Simulation*, 60(3-5):169–180, 2002.

116. J.-J. Slotine and W. Li. *Applied nonlinear control*. Prentice Hall, 1991.

117. M. Staroswiecki. Fault tolerant control: the pseudo-inverse method revisited. In *16th IFAC World Congress*, Prague, Czech Republic, 2005.

118. A. Stríbrský, K. Hyniová, J. Honcu, and A. Kruczek. Using fuzzy logic to control active suspension system for one half-car-model. *Acta Montanistica Slovaca*, 8(4):223–227, 2003.

119. K.J. Åström and R.M. Murray. *Feedback systems: An introduction for scientists and engineers*. Princeton University Press, 2008.

120. D. Theilliol, C. Join, and Y. Zhang. Actuator fault-tolerant control design based on reconfigurable reference input. *International Journal of Applied Mathematics and Computer Science - Issues in Fault Diagnosis and Fault Tolerant Control*, 18(4):555–560, 2008.

121. D. Theilliol, H. Noura, and J.-C. Ponsart. Fault diagnosis and accommodation of a three-tank-system based on analytical redundancy. *ISA Transactions*, 41(3):365–382, 2002.

122. D. Theilliol, J.-C. Ponsart, M. Mahfouf, and D. Sauter. Active fault tolerant control design for an experimental hot rolling mill - a case study. In *12th IFAC Symposium on Automation in Mining, Mineral and Metal Processing*, pages 113–118, Quebec City, Canada, 2007.

123. D. Theilliol, J.-C. Ponsart, M. Rodrigues, S. Aberkane, and J. Yamé. Design of sensor fault diagnosis method for nonlinear systems described by linear polynomial matrices formulation: application to a winding machine. In *17th IFAC World Congress*, pages 1890–1895, Seoul, Korea, 2008.

124. D. Theilliol, J.-C. Ponsart, D. Sauter, M. Mahfouf, and M.A. Gama. Design of a fault diagnosis system based on a bank of filter-observers with application to a hot rolling mill. *Transactions of the Institute of Measurement and Control*, 2009. to appear.

125. H.L. Trentelman, A.A. Stoorvogel, and M. Hautus. *Control theory for linear systems*. Springer, Series: Communications and Control Engineering, 2001.

126. V. Venkatasubramanian, R. Rengaswamy, and S.N. Kavuri. A review of process fault detection and diagnosis. part ii: Qualitative models-based methods. *Computers and Chemical Engineering*, 27(3):313–326, 2003.

127. V. Venkatasubramanian, R. Rengaswamy, S.N. Kavuri, and K. Yin. A review of process fault detection and diagnosis. part iii: Process history based methods. *Computers and Chemical Engineering*, 27(3):327–346, 2003.

128. V. Venkatasubramanian, R. Rengaswamy, K. Yin, and S.N. Kavuri. A review of process fault detection and diagnosis. part i: Quantitative model-based methods. *Computers and Chemical Engineering*, 27(3):293–311, 2003.

129. J. Wang, A.C. Zolotas, and D.A. Wilson. Active suspensions: a reduced-order H_∞ control design study. In *Mediterranean Conference on Control and Automation*, pages 1–7, Athens, Greece, 2007.

130. R.L. Williams-II and D.A. Lawrence. *Linear state-space control systems*. John Wiley & Sons, Inc., 2007.

131. E.N. Wu, S. Thavamani, Y. Zhang, and M. Blanke. Sensor fault masking of a ship propulsion. *Control Engineering Practice*, 14(11):1337–1345, 2006.

132. H. Xu and P.A. Iaonnou. Adaptive sliding mode control design for a hypersonic flight vehicle. CATT technical report No. 02-02-01, 2001.

133. D. Xue, , Y. Chen, and D.P. Atherton. *Linear feedback control: Analysis and design with* MATLAB *(Advances in design and control)*. Society for Industrial Mathematics, first edition, 2008.

134. N. Yagiz. Comparision and evaluation of different control strategies on a full vehicle model with passenger seat using sliding modes. *International Journal of Vehicle Design*, 34(2):168–182, 2004.

135. N. Yagiz and I. Yuksek. Sliding mode control of active suspensions for a full vehicle model. *International Journal of Vehicle Design*, 26(2-3):265–276, 2001.

136. N. Yagiz, I. Yuksek, and S. Sivrioglu. Robust control of active suspensions for a full vehicle model using sliding mode control. *JSME International Journal. Series C: Mechanical Systems, Machine Elements and Manufacturing*, 43(2):253–258, 2000.

137. Z. Yang, R. Izadi-Zamanabadi, and M. Blanke. On-line multiple-model based adaptive control reconfiguration for a class of nonlinear control systems. In

4th IFAC Symposium on Fault Detection, Supervision and Safety for Technical Processes, pages 745–750, Budapest, Hungary, June 2000.

138. Q. Zhang, M. Basseville, and A. Benveniste. Fault detection and isolation in nonlinear dynamic systems: a combined input-output and local approach. *Automatica*, 34(10):1359–1373, 1998.

139. Y. Zhang and J. Jiang. An interacting multiple-model based fault detection, diagnosis and fault-tolerant control approach. In *Proceedings of the IEEE American Control Conference*, pages 3593–3598, Phoenix, AZ, USA, 1999.

140. Y. Zhang and J. Jiang. Bibliographical review on reconfigurable fault-tolerant control systems. *Annual Reviews in Control*, 32(2):229–252, 2008.

141. Y.M. Zhang and J. Jiang. Issues on integration of fault diagnosis and reconfigurable control in active fault-tolerant control systems. In *6th IFAC Symposium on Fault Detection, Supervision and Safety for Technical Processes*, pages 1513–1524, Beijing, P.R. China, 2006.

142. D.H. Zhou and P.M. Frank. Nonlinear adaptive observer based component fault diagnosis of nonlinear in closed-loops. In *14th IFAC World Congress*, pages 25–30, Beijing, P.R. China, 1999.

143. D.H. Zhou, G.Z. Wang, and S.X. Ding. Sensor fault tolerant control of nonlinear systems with application to a three-tank-systems. In *4th IFAC Symposium on Fault Detection, Supervision and Safety for Technical Processes*, pages 810–815, Budapest, Hungary, 2000.

Index

Other titles published in this series (continued):